MATHEMATICAL INTUITION

VOLUME 203

Library of Congress Cataloging in Publication Data

Tieszen, Richard.
Mathematical intuition : phenomenology and mathematical knowledge
by Richard Tieszen.
p. cm. -- (Synthese library ; v. 203)
Bibliography: p.
Includes index.
ISBN 0-7923-0131-5
1. Mathematics--Philosophy. 2. Intuition. I. Title.
II. Series.
QA8.4.T53 1989
510'.1--dc19 88-37514

ISBN 0-7923-0131-5

Published by Kluwer Academic Publishers,
P.O. Box 17, 3300 AA Dordrecht, The Netherlands.

Kluwer Academic Publishers incorporates
the publishing programmes of
D. Reidel, Martinus Nijhoff, Dr W. Junk and MTP Press.

Sold and distributed in the U.S.A. and Canada
by Kluwer Academic Publishers Group,
101 Philip Drive, Norwell, MA 02061, U.S.A.

In all other countries, sold and distributed
by Kluwer Academic Publishers,
P.O. Box 322, 3300 AH Dordrecht, The Netherlands.

printed on acid free paper

Printed in The Netherlands

RICHARD L. TIESZEN

MATHEMATICAL INTUITION

Phenomenology and Mathematical Knowledge

KLUWER ACADEMIC PUBLISHERS

DORDRECHT / BOSTON / LONDON

RICHARD L. TIESZEN

MATHEMATICAL INTUITION

Phenomenology and Mathematical Knowledge

KLUWER ACADEMIC PUBLISHERS

DORDRECHT / BOSTON / LONDON

Library of Congress Cataloging in Publication Data

Tieszen, Richard.
Mathematical intuition : phenomenology and mathematical knowledge
by Richard Tieszen.
p. cm. -- (Synthese library ; v. 203)
Bibliography: p.
Includes index.
ISBN 0-7923-0131-5
1. Mathematics--Philosophy. 2. Intuition. I. Title.
II. Series.
QA8.4.T53 1989
510'.1--dc19 88-37514

ISBN 0–7923–0131–5

Published by Kluwer Academic Publishers,
P.O. Box 17, 3300 AA Dordrecht, The Netherlands.

Kluwer Academic Publishers incorporates
the publishing programmes of
D. Reidel, Martinus Nijhoff, Dr W. Junk and MTP Press.

Sold and distributed in the U.S.A. and Canada
by Kluwer Academic Publishers Group,
101 Philip Drive, Norwell, MA 02061, U.S.A.

In all other countries, sold and distributed
by Kluwer Academic Publishers,
P.O. Box 322, 3300 AH Dordrecht, The Netherlands.

printed on acid free paper

Printed in The Netherlands

For Nancy

TABLE OF CONTENTS

PREFACE

"Intuition" has perhaps been the least understood and the most abused term in philosophy. It is often the term used when one has no plausible explanation for the source of a given belief or opinion. According to some sceptics, it is understood only in terms of what it is not, and it is not any of the better understood means for acquiring knowledge. In mathematics the term has also unfortunately been used in this way. Thus, intuition is sometimes portrayed as if it were the Third Eye, something only mathematical "mystics", like Ramanujan, possess. In mathematics the notion has also been used in a host of other senses: by "intuitive" one might mean informal, or non-rigourous, or visual, or holistic, or incomplete, or perhaps even convincing in spite of lack of proof.

My aim in this book is to sweep all of this aside, to argue that there is a perfectly coherent, philosophically respectable notion of mathematical intuition according to which intuition is a condition necessary for mathematical knowledge. I shall argue that mathematical intuition is not any special or mysterious kind of faculty, and that it is possible to make progress in the philosophical analysis of this notion.

This kind of undertaking has a precedent in the philosophy of Kant. While I shall be mostly developing ideas about intuition due to Edmund Husserl there will be a kind of Kantian argument underlying the entire book. The style of argument is clearly present in Husserl's later philosophy. Simply put, it is to ask how mathematical knowledge is possible; in particular, to ask how knowledge of number or of other mathematical objects is possible. In Husserl's philosophy this is an especially interesting question, for Husserl wanted to be a kind of platonist or realist about mathematical objects, and yet to develop an epistemology that was compatible with being a realist. To succeed in any measure at this task would be to make progress with one of the main problems in the philosophy of mathematics.

The main positive argument of the book can be looked at as comprised of two subarguments: first, arguments to show that mathematics is about abstract objects, objects that are eternal, unchanging, causally inert, and outside of spacetime. Husserl gives arguments to the effect that sense experience does not contain abstract mathematical objects, and that mathematics has a kind of universality and necessity not found in sense experience.

I discuss these arguments, and I think there is much to be said for them, but in this book I do not pursue them as far as I think they could be pursued. Secondly, I consider arguments that purport to show how it is possible to have knowledge about such objects, where knowledge requires intuition, based on the Kantian strategy. Thus, the idea is to determine the kinds of cognitive structures and processes which are neccesary for knowledge about numbers and finite sets, where these objects are understood as abstract. This part of the argument is taken up in some detail and the consequences are explored for a variety of major issues in the philosophy of mathematics.

Chapter 1 starts with a brief statement of how the notions of knowledge, evidence and intuition are connected and then surveys recent views on mathematical intuition. It is interesting to look at these views as more or less adequate attempts to answer the Kantian question. The main problems about mathematical intuition that I would like to address in the book are then reviewed: problems about the analogousness of mathematical intuition and perception, about mathematical structuralism, and about the sense in which mathematical knowledge is supposed to be "founded" on mathematical intuition.

In Chapter 2 I turn to Husserl's writings to describe the phenomenological conception of intuition. In deference to Husserl scholarship I have quoted extensively from the texts to establish my interpretation on this issue. At the center of the phenomenological view is the concept of intentionality, and it is by approaching intuition via intentionality that I believe it is possible to obtain new insights into mathematical intuition and mathematical knowledge. Chapter 3 is devoted to an examination of perceptual intuition and is intended to set the stage for the analogies I wish to draw with mathematical intuition later in the book.

Chapter 4 gives a detailed statement of objections to the notion of mathematical intuition, along with some preliminary indications about how they will be met. Toward the end of the chapter I present the framework for the account of mathematical intuition to follow. The main idea here is to set up logic for different domains of objects in such a way that it does not abstract from intuitability conditions. Intuition is understood in terms of fulfillment (or fulfillability) of intentional acts, and following Heyting, Martin-Löf and others, I say that an intentional act whose content is expressed by S is fulfilled (or is fulfillable) if and only if there is a construction for (or proof of) S. On our view, a theory of constructions, suitably understood, *is* a theory of mathematical intuition.

Chapters 5 and 6 work through many of the details of this kind of view in

the case of natural numbers. Thus, questions about definability of number, the origin of the concept of number, small and large numbers, and mathematical induction are looked into, and objections to this view of mathematical intuition are parried one by one. Chapter 7 takes a similar approach to finite sets, where finite sets are viewed as objects obtained from given objects (initially individuals) by construction processes.

Chapter 8 briefly summarizes the phenomenological approach and indicates how it could be used to deal in a new way with a variety of issues, including issues about anti-realism and realism, and mathematical truth and knowledge.

Although the arguments of later chapters often depend on the arguments of earlier chapters, there are different ways to read this book. The book could be read, for example, as a detailed development of a philosophically defensible view of the the interpretation of the intuitionistic logical constants, especially in connection with arithmetic. This way of looking at the work in the book results from Heyting's interpretation of the intuitionistic logical constants in terms of a theory of intentionality (in Husserl's sense), an interpretation which is described in Chapter 4.

In thinking about the matters discussed in the book I have benefited from the comments, suggestions, and criticisms of a number of people. I profited from many conversations with Charles Parsons on the foundations of mathematics and on phenomenology while I was a doctoral student at Columbia University. In addition to offering comments on earlier versions of this essay he kindly made available some of his own unpublished work-in-progress on the topic of mathematical intuition, which helped me to put my ideas into clearer focus. I am greatly indebted to him. I also benefited from comments by a number of other philosophers at Columbia: Isaac Levi, Shaughan Lavine, and Paul Rosenbloom.

I thank Wilfried Sieg for encouragement and for comments at an earlier stage of the work, especially on matters concerning constructive mathematics. Robert Tragesser offered a number of stimulating observations and I have enjoyed several discussions with him on phenomenology and mathematics. I also received many valuable comments on the manuscript from Michael Resnik. John Ackermann deserves thanks for originally bringing this to pass. Were it not for Michael Resnik and the Summer Seminar program of the National Endowment for the Humanities I would not have had the time to bring the project to completion.

Several other people read parts of earlier versions of this work and offered important comments and suggestions. I thank J.N. Mohanty, my principal

teacher in phenomenology at the New School for Social Research, and Dagfinn Føllesdal, whose writings on phenomenology have been a source of inspiration, for their remarks on Husserl and phenomenology. I am grateful to Hao Wang for encouragement and for discussion on set theory and Gödel.

The comments of several anonymous referees at *Nous* and *Synthese* on papers I published in those journals were helpful at an early stage in the development of some of the ideas in the book. I am also grateful to Dr. Corliss Lamont who provided a grant to the Department of Philosophy at Columbia which made it possible to begin the research that led to this book.

Finally, I am indebted to Karen Holt and Paul Shields for help with the original production of the manuscript on computer.

ABBREVIATIONS FOR WORKS OF HUSSERL

CM	Cartesian Meditations
EJ	Experience and Judgment
FTL	Formal and Transcendental Logic
Ideas	Ideas: General Introduction to Pure Phenomenology
LI	Logical Investigations
PA	Philosophie der Arithmetik
PITC	The Phenomenology of Internal Time-consciousness

CHAPTER 1

THE CONCEPT OF INTUITION IN MATHEMATICS

1. INTRODUCTION

The notion of mathematical intuition has played an important role in several philosophical conceptions of how we acquire mathematical knowledge and a number of classical philosophical problems are closely associated with it: the question whether mathematical knowledge is a priori, analytic or "necessary"; problems about platonism, mathematical truth, and abstract objects; the question whether minds surpass machines, or whether mathematics consists solely of mechanical operations on syntax; questions about the meaning and reference of statements of mathematical theories; and questions about the nature of infinity, to name a few.

In this essay I shall discuss the role of intuition in the foundations of mathematics. The concept of mathematical intuition has been subject to a variety of interpretations and developments and the most important recent developments will be surveyed in this chapter. My own approach will be to use phenomenological considerations to try to come to a clearer understanding of the role of intuition in mathematical knowledge.

In speaking of phenomenological considerations I shall have in mind the ideas of Husserl. Husserl's view of knowledge derives from the Kantian tradition in which knowledge is thought to be a product of what Kant calls concepts and intuitions. In Husserl's writings one has a similar distinction between (empty) intentions and intuitions, where intuition (Anschauung) is understood as the "fulfillment" of intention. It is my conviction that the development of Husserl's view in the context of mathematics is not merely of historical interest, but that it would contain ideas quite relevant to contemporary issues in the philosophy and foundations of mathematics. Much has of course transpired since Husserl's time and many issues with which Husserl dealt have been greatly refined or even completely reshaped by later events. There are also many difficulties in understanding Husserl's own views due to some obscurities in his writing and in particular to a lack of detail about the notion of *mathematical* intuition. In this study I shall analyze and develop a number of Husserl's ideas in light of contemporary issues in the philosophy and foundations of mathematics with a view to giving an account of mathe-

matical intuition that will meet some of the well-known objections that have been raised to this notion. These objections will first be mentioned in the final section of this chapter and then discussed in more detail in Chapter 4. Responses to the objections are provided in Chapters 4 through 8. The purpose of this chapter is to discuss some of the main issues associated with the notion of mathematical intuition and to briefly survey some of the better known views on mathematical intuition.

2. KNOWLEDGE, EVIDENCE, AND INTUITION

My discussion of the role of intuition in mathematical knowledge will not proceed as if we have in advance an account of what knowledge is. In order to fix ideas, however, let us start with the view that a necessary condition for a mathematician M to *know* that S is that M believe S, and that M's belief that S be produced by a process which gives evidence for the belief. We shall be discussing these kinds of conditions throughout this essay. We might note immediately that on the view we shall be discussing evidence is provided by *intuition*. Thus when we say that M has evidence for S, we mean that M sees that S is the case in a direct and noninferential way. Intuition provides something like direct acquaintance. The "evidence" provided by intuition is therefore to be distinguished from "justification" insofar as the latter is thought to provide knowledge by way of inference or by way of beliefs that are supposed to support other beliefs, as in the case of indirect knowledge. The relevant notion of evidence is introduced and discussed in Chapter 2.

Suppose we add to the above conditions the condition that S be true. Then would these three conditions be sufficient for mathematical knowledge? The question can be left open at this point. There are some indications in Husserl's writings that, provided the evidence is of the proper kind (e.g., adequate, apodictic), these conditions would be sufficient. For example, let S be Fermat's last theorem $(\forall n)Fn$ and suppose M is a mathematician who believes he has found an exponent m such that $\neg Fm$. Suppose further that $\neg Fm$ is true, and that M's belief that $\neg Fm$ is produced by a process which gives evidence for it. It could perhaps happen under these circumstances that we would not want to say that M knows that $\neg Fm$. However, if we begin to place constraints on the kind of evidence that would be required – for example, that it be adequate or apodictic in Husserl's sense – it seems fairly evident that our conditions would be sufficient. It is another question, however, about which I shall have something to say in this essay, whether we could ever hope to have adequate or apodictic evidence for our mathematical

beliefs. The notions of adequacy and apodicticity are also discussed in Chapter 2.

It should be noted that in our conditions for knowledge we do not suppose that the state of affairs referred to by the statement S *causes* M to believe S. One of the difficult questions about mathematical knowledge has been to explain how or whether the state of affairs referred to by S could possibly *cause* a person to believe S. Examples from mathematics, like our example of Fermat's last theorem, will quickly give reason to pause over this question. Indeed it might seem absurd that there could be anything like causal relations involved in knowledge about mathematical objects. Fermat's last theorem, for example, is evidently about ordered quadruples $\langle x, y, z, n\rangle$ of positive integers with $n > 2$, and unlike the case of perceptual objects it is far from clear how there could be causal relationships with such objects. Of course one might suppose that mathematical statements could be systematically reconstrued so that they do not refer to abstract objects like numbers, sets or functions, but in spite of some of the best recent work in this direction there are as many difficulties with carrying out this program as there are with eschewing causal accounts of knowledge. A phenomenological elucidation of knowledge will allow us to avoid this tangle from the outset. For a causal account of knowledge is not part of the subject matter of phenomenology even in the case of ordinary perception. It might be that considerations about causality would be necessary to supplement a phenomenological account but such matters could be taken up later.

I have said that it is a condition for knowledge that M's belief that S be produced by a process which gives evidence for the belief. The role of intuition in mathematical knowledge is to provide the relevant evidence. I shall concentrate in particular on the evidence provided for the *existence* of mathematical objects of two types natural numbers and hereditarily finite sets but I shall later say something about how the view I develop extends to other parts of mathematics. The role of intuition in mathematics will thus be quite similar to the role of perceptual intuition in providing evidence for the existence of physical objects. Anyone sympathetic to a Kantian view of knowledge would naturally expect that intuition should have some such role as this in mathematical knowledge. In fact one could even point to earlier views of this type in the history of philosophy. The recent history of logic and the foundations of mathematics can be understood as giving some impetus to this view. It could be argued that one of the more significant developments in this regard is the failure of Frege and Russell to derive the principles of number from logic. This failure might be taken to show that the principles of

number are extra-logical. At least one way in which they are extra-logical is with respect to their existence assumptions. That numbers or sets should exist or at least be possible in some way cannot be made out, so it appears, on the basis of sheer logical reasoning. But then how shall we explain our capacity to know truths about them? It is a very natural suggestion that we know truths about numbers and sets in a way analogous to the way that we know truths about physical objects: by some kind of intuition.

Since I want to concentrate on questions about the intuition of natural numbers and finite sets I shall not have much to say about geometric intuition. There are interesting questions that could be raised about geometric intuition and, in particular, about how it might be related to number-theoretic or set-theoretic intuition.[1] One might, for example, view *formal theories* of numbers and finite sets, and formal systems in general as a kind of geometry of signs. But surely the intuition of signs, whether as types or tokens, could not give knowledge about numbers or sets. The intuition of a string of symbols like Osssssssss is not the same as the intuition of a particular number, although we often use signs to represent numbers. I shall argue that our knowledge of numbers and sets could not be founded on what we can learn about signs and properties of signs in ordinary sense-perception. The notion of "founding" has, however, been used in different ways in the literature and we shall need to be careful about distinguishing these differences. Charles Parsons, for example, has offered a view of mathematical intuition in which the intuition of abstract objects, sign types, is said to be founded on perceptual intuition. He does not, however, suggest that what we can learn by examining signs and their properties would lead us to knowledge about number. What could evidently be said is that this kind of intuition will at least guarantee the existence of the elements of a structure satisfying the principles of arithmetic. I shall come back to his views momentarily.

We shall be speaking of knowledge as having, in a special sense, a "foundation" in intuition but we should perhaps dispel from the outset some of the "foundationalist" overtones that this may carry.[2] The view we shall develop, based on Husserl's later ideas, will not at all imply that intuition provides an absolutely certain foundation for our knowledge. On the contrary, what is given in intuition is subject to error and revision, and this will be true for both perceptual and mathematical intuition. The evidence provided by intuition in either case comes in different types and degrees.

3. INTUITION "OF" AND INTUITION "THAT"

In our discussion of a mathematician's knowledge that S we should distinguish between intuition *of objects* and intuition *that* something is the case.[3] The distinction marks the difference between the case where a mathematician M intuits x, where we take "x" to refer to some object (i.e., substitute for "x" a singular term), and the case where M intuits that S, where "S" refers to some state of affairs (i.e., substitute for "S" a sentence). I shall mainly investigate intuition *of* particular objects – natural numbers and HF sets. The idea that there is intuition *of objects* in mathematics has been involved in several conceptions of the foundations of mathematics. Some constructivist views, like those of Hilbert and Bernays on "contentual" or "finitist" mathematics, have clearly invoked this idea. For Hilbert and Bernays the objects of intuition were to be finite configurations of symbols, perhaps symbol *types*. For other constructivists like Brouwer or Weyl it is not so clear whether intuition is to be understood as intuition *of* particular objects. But there seems to also be something like a concept of intuition *of* numbers and sets in Gödel's writings on set theory, which would normally be thought of as the paradigm of a platonistic theory.

While I shall concentrate mainly on intuition *of*, my account will also be concerned with intuition *that* and with the relation between intuition *that* and intuition *of*. I think that some aspects of the relationship will emerge in a fairly clear way in the specific contexts of number theory and finite set theory below. As I shall understand the matter, intuition *of* will typically be necessary for M to know *that* S. Knowledge that something is the case, as it is built up genetically and becomes part of our background knowledge, can also affect just what intuition is *of*. The role of background beliefs in making intentions determinate is an important part of the phenomenology of both perceptual and mathematical knowledge. As we acquire knowledge our intentions become more determinate. Background beliefs will function to make *what* we intuit more determinate.

Another closely related distinction has been made between *de re* and *de dicto* intuition.[4] From the point of view of phenomenology the fulfillment of *de re* intentions would amount to *de re* intuition, while the fulfillment of *de dicto* intentions would amount to intuition *de dicto*. The distinction between *de re* and *de dicto* is usually made for propositional attitudes. It is not so clear how to make it for singular terms although it might be useful to do so. We should be able to say that an intention to a in the context of the proposition that Fa is *de re* iff the intention to the state of affairs that Fa is *de re*. I shall

not be analyzing the distinction in any detail in this study but I think it would be worthwhile to do so, especially in connection with the singular terms of mathematical theories that have the character of proper names.

4. SOME RECENT VIEWS OF MATHEMATICAL INTUITION

Let us now briefly survey some of the most important views on mathematical intuition that have been developed in recent times. In several cases these views have been directly related to the philosophy of Kant. In the philosophy of Kant one finds an important role assigned to intuition in mathematics. Kant of course appealed to the notion of intuition to answer the question of how mathematical knowledge was possible. Just how the details of Kant's views on mathematical intuition are to be understood has been the source of a substantial amount of interpretation and analysis, and it is not possible to do justice to any of it here.[5] Kant's work is in part so significant because it places issues connected with the notion of mathematical intuition in the context of many wider philosophical problems. I shall not attempt to describe Kant's views here at all but I shall occasionally discuss some affinities and differences between his views and phenomenological views.

The notion of mathematical intuition has been interpreted in several different ways in recent work. For *formalists* or *finitists* who follow Hilbert and Bernays intuition is supposed to be *of* finite sign configurations. This could either mean that intuition is of sign tokens or of sign types and one would have to develop some view about the matter. For traditional *intuitionists* like Brouwer and Heyting it is clear that intuition is not supposed to be of finite sign configurations. Rather, it is described as a mental process that one carries out. It is not clear on this view whether there is supposed to be intuition *of* objects at all, or whether there is supposed to be intuition of something like the process of iteration. Brouwer sometimes speaks of intuition as if it were of mental entities or processes, and then tends to assimilate the notion of intuition to introspection. Finally, *platonists* like Gödel who invoke the notion of intuition take it to be of abstract objects like numbers, sets, groups and so on, not of sign configurations, or mental entities. Let us briefly consider each kind of view.

5. HILBERT AND BERNAYS

Hilbert's views on intuition have been widely discussed and analyzed. It is well-known that Hilbert, like Brouwer, actually invoked Kant's views in

developing his ideas on the role of intuition in the foundations of mathematics. One of Hilbert's well-known statements is still worth quoting in full for the conception it presents of the role of mathematical intuition in mathematical knowledge:

> Kant already taught ... that mathematics has at its disposal a content secured independently of all logic and hence can never be provided with a foundation by means of logic alone; that is why the efforts of Frege and Dedekind were bound to fail. Rather, as a condition for the use of logical inferences and the performance of logical operations, something must already be given to our faculty of representation (in der Vorstellung), certain extra-logical concrete objects that are intuitively (anschaulich) present as immediate experience prior to all thought. If logical inference is to be reliable, it must be possible to survey these objects completely in all their parts, and the fact that they occur, that they differ from one another, and that they follow each other, or are concatenated, is immediately given intuitively, together with the objects, as something that can neither be reduced to anything else nor requires reduction. This is the basic philosophical position that I consider requisite for mathematics and, in general, for all scientific thinking, understanding, and communication. And in mathematics, in particular, what we consider is the concrete signs themselves, whose shape, according to the conception we have adopted, is immediately clear and recognizable.[6]

It is clear from these and other remarks in Hilbert's writing that the concrete *objects* of intuition were to be finite configurations of signs. It is questionable, however, whether Hilbert meant that sign tokens, like physical inscriptions of strokes, were to be the objects of intuition, or rather whether the objects of intuition should be thought of as sign types. In the latter case it would not be possible to suppose that the objects could be given ordinary sense-perception although it would be possible to argue that the intuition of such objects is at least founded on sense-perception. But the distinction already raises a host of issues about mathematical intuition that Hilbert simply does not address. Hilbert no doubt knew of Kant's idea of pure intuition but his own views on mathematical intuition are not very well developed.

Hilbert's views on the intuitive content of proofs by mathematical induction have been construed in different ways. They seem to at least entail that *instances* of induction are verifiable in intuition provided that the predicates involved are of an elementary nature, e.g., primitive recursive. For

then an *instance* can be obtained by uniformly substituting whatever signs are playing the role of numerals for variables in the induction schema and then verifying that the predicates hold. Verifying that the predicates hold will involve some elementary procedure like computing a primitive recursive function.

A number of problems can be raised about Hilbert's view of the intuitive content of proofs by mathematical induction. A problem with which I shall be concerned is whether we can see by intuition that induction as a *general* principle holds. Interpreters of Hilbert have taken different positions on Hilbert's view of this matter. Kreisel, for example, has argued that Hilbert did take some general (universally quantified) statements about natural numbers to be part of finitist mathematics. Unless one then jettisons Hilbert's concern for intuition in mathematical knowledge there would presumably be some intuitive capacities which allow us to see that such statements hold. This question will occupy us in a moment.

Bernays was more actively engaged in working out a philosophical conception of mathematical intuition which would underlie Hilbert's conception of proof theory and which would be based on ideas of Kant, as is shown by a number of the essays now collected as *Abhandlungen zur Philosophie der Mathematik*, and by the early sections of the *Grundlagen der Mathematik I*. In light of Gödel's incompleteness theorems Bernays began to see the need for revising his epistemological ideas. As Gödel later put it, it would be necessary to drop the original finitist restriction to concrete mathematical intuition in order to give consistency proofs for elementary arithmetic and subsystems of analysis. Some of the epistemological proposals that accompanied this realization will be surveyed below.

6. PARSONS

The most thoroughly developed ideas in recent times on what might be considered a modified formalist or finitist view of mathematical intuition are those due to Charles Parsons. It will not be possible to do justice to Parsons' views on intuition in this survey, for they are connected with many other themes in his work that we cannot go into here, eg., Kant, structuralism, quasi-concrete objects, modal notions, substitutional quantification, etc. Nonetheless, it will be useful for our discussion later to briefly survey some of the ideas on mathematical intuition he has offered in his writings.[7] Parsons starts with something like a Hilbertian conception of mathematical intuition according to which the objects of intuition are finite configurations of signs,

like strings of strokes. On the basis of Parsons' phenomenology of strings of strokes one is permitted to say that there is intuition of sign *types* so long as concrete instances, sign tokens, can be perceived or at least imagined to be in spacetime. Sign types are "quasi-concrete" abstract objects in the sense that they are determined by intrinsic relations to concrete objects, sign tokens. What object a type is, is determined by what are or would be its tokens. The type is, so to speak, manifested in the perceived token. Sign tokens are as unquestionably available to ordinary sense-perception as anything could be. At least in small finite cases they admit of ostension, they are "complete", and they are evidently causally related to our senses. Parsons makes some suggestions about large finite cases and induction by explicitly appealing to modal notions about sign configurations that *could* be constructed. He points out that stroke-string types are minimally abstract kinds of objects. For these objects there is a clear sense in which we can say that the intuition of the object is founded on sense-experience or on imaginings of objects as being in space and time.

Parsons notes that his account may not take us beyond elementary syntax and perhaps traditional geometry. But what kind of role might intuition have in substantial mathematical theories about numbers and sets? Could natural numbers or finite sets be objects of intuition? On the basis of the conception of intuition just developed it is perhaps best to take the view that intuition gives objects, like strings of strikes, which can be taken as the elements of a model of arithmetic. On this view, intuition could be taken to guarantee the existence of the elements of a structure of the sort required to satisfy principles of elementary number theory. But we should not say that "the" natural numbers are themselves objects of intuition since intuition in this sense does not give a unique sequence to be "the" natural numbers, and since the concept of number does not rule out as the intended model objects that are not objects of intuition in this sense. This kind of account of mathematical intuition, Parsons points out, would be compatible with a structuralist view of number for the simple reason that on it there is no intuition of numbers. It does not attempt to construe numbers as any particular objects. The account nonetheless shows how mathematical intuition plays a role in founding mathematical knowledge.

Parsons' view gives a sense in which one can say that knowledge of natural numbers is *founded* on ordinary perceptual intuition. Intuition gives objects which can be taken as elements of a model of arithmetic. Hence, intuition could be taken to guarantee the existence of elements of a structure of the sort required to satisfy principles of arithmetic. We might say that

intuition certifies that a structure about which we have certain beliefs *exists*, or at least that its elements exist. It might be held that apart from intuition we could not have this kind of assurance. This view of "founding", it should be noted, is quite different from any conception of founding according to which we come to intuit numbers and to understand their properties solely on the basis of intuition of sign configurations, for on the view suggested by Parsons there is no intuition of numbers. The alternative view of founding has been imputed to philosophers like Kant and Hilbert by, for example, Kitcher, although I am not sure that they actually held it.[8]

Parsons also explores in detail the view that natural numbers and finite sets could be objects of intuition. For natural numbers he develops two main proposals.[9] The first proposal amounts to thinking of natural numbers as generalized types. Here one thinks of numerals of different systems as "tokens" of a type, where the relation of sameness of type is that of "representing the same number". Numbers as thus understood are a kind of generalized type since the relation of sameness of type is more abstract than on the earlier conception. The second proposal sees numbers as "constituted" by language in a sense given by a substitutional interpretation of quantifiers in the language of arithmetic. Parsons considers many aspects of each proposal that cannot be gone into here. But a general conclusion drawn about both views is that they are particular *construals* of the natural numbers and hence go beyond a merely structural characterization of what the natural numbers are. They both attach to the concept of number something that is inessential to it – a kind of perceptual intuition founded on signs or expressions. In this sense they run afoul of structuralism. Parsons points out that when it comes to guaranteeing the existence of a structure of the sort needed to satisfy principles of number theory, neither conception gives more than the earlier proposal of intuition of sign types. In spite of this fact it is suggested that we might think of the substitutional or generalized types conception as having a role in a genetic account of mathematical language which could shed light on how numbers come into our ontology.

About the intuition of finite sets Parsons also has much to say. For a given domain D of individuals the HF sets built from D, HF(D), are obtained by beginning with the empty set Ø and iterating the operations of forming from any $x_1 \ldots x_n$, individuals or previously given sets, the set $\{x_1 \ldots x_n\}$ whose elements are just $x_1 \ldots x_n$. Now the idea developed by Parsons is that if D consists of possible objects of intuition then we can regard HF(D) as constructible in intuition in roughly the same sense in which strings of strokes are. The proposal that Parsons explores at some length involves

understanding the language of the theory of HF sets in terms of a relative substitutional semantics. In this context Parsons again discusses the thesis that the mathematical objects, in this case finite sets, are constituted by language. In the case of HF sets the idea would be that we can view such sets as constituted by language and by the individuals of an underlying domain from which HF sets are constituted. The general assessment of this proposal is that it gives a natural description of the minimal commitments of someone who speaks of HF sets with ultimate elements arbitrarily chosen from a domain over which he quantifies objectually. But it again adds something extraneous to our concept of set, namely that the individuals of the underlying domain must be perceptually intuitable in a particular way. Moreover, it is not clear what we could make of such a view for sets in general. Insofar as we view finite sets as instances of sets in general the substitutional semantics will not capture one aspect of the concept of finite set. Parsons again suggests that we might see the substitutional semantics for finite sets as representing a stage in the genesis of the concept of set, somewhat in the spirit of Quine's view in the *Roots of Reference*.

Parsons also discusses questions about the intuitability of transfinite sets, particularly in connection with the (maximal) iterative conception of set.[10] We shall not take up any of these views except perhaps to occasionally mention a view that would be of relevance to a point under discussion.

On the question of the intuitive character of mathematical induction Parsons distinguishes between intuition of objects and the kind of evidence we have for mathematical induction and other principles of the same kind. In a paper on Kant in which he discusses the matter of induction a fruitful distinction of this kind is made:

> There seem to be two independent types of insight into our forms of intuition which a Kantian view requires us to have, that which allows a particular perception to function as a "formal intuition" and that which we have into the *possible progression* of the generation of intuitions according to a rule. To speak of a peculiar *kind* of intuition in the second case seems quite misleading. The mathematical knowledge involved has a highly complex relation to "intuition" in the more specifically Kantian case.[11]

The evidence we have for mathematical induction could not be due to a kind of intuition giving *objects*, since it concerns general propositions about objects that could be objects of intuition in the first sense. Parsons suggests that on the basis of reflections of Bernays and Gödel on the difference

between the kind of evidence Hilbert claimed for finitist mathematics and constructive evidence generally it follows that induction, even applied to objects of intuition like strings of strokes, is *not* intuitively evident. This would be true for a Hilbertian conception of concrete mathematical intuition because induction involves "abstract" notions like that of predicate, property or class. Of Hilbert's idea that at least individual instances of induction are intuitively evident where predicates are, say, primitive recursive, Parsons argues that there is still something involved here that is not involved in our evidence for singular statements. What is involved is understanding of a predicate such as "is a string (from a given alphabet)" which gives the domain of the general statement. Hence, Hilbert's claim is doubtful, but even if accepted the explanation of the intuitive evidence involved would have to be different from that in the singular case. Parsons suggests that a Hilbertian could try to view induction as something like a logical principle, and then point out that it is *conservative* of intuitive evidence: it leads from intuitively evident premises to intuitively evident conclusions.[12]

7. BROUWER

Another account of mathematical intuition which tries to incorporate the insight into mathematical induction more directly is due to Brouwer. Brouwer's remarks on mathematical intuition are not developed in much detail but it is clear that what Brouwer took to be "intuitive" in mathematics went beyond what could be justified on the basis of Hilbert's original conception of intuition. Brouwer's idea of intuition is that there is an awareness of the movement of time which is "the falling apart of a life moment into two distinct things, one which gives way to the other, but is retained by memory."[13] If we abstract from the qualitative elements of this "two-ity" there remains the "common substratum of all two-ities", which we might think of as the form of two-ity. Brouwer says that the awareness of this basic two-oneness is the "original intuition of mathematics". It is the foundation of the natural numbers inasmuch as one of the elements of the two-oneness may be thought of as new two-oneness, and this *process* may be iterated indefinitely. This view of intuition is taken by intuitionists to justify proofs by mathematical induction in a fairly straightforward way, so that it is reasonable to claim in intuitionistic mathematics that we know statements proved by mathematical induction hold, not only that individual instances of such statements can be verified in intuition.

As I mentioned earlier, it might be argued that on Brouwer's conception of

the "original intuition of mathematics" we do not have a view about intuition *of objects*, in particular, of natural numbers, nor do we have the view that intuition is *of* iteration. The view that there is actually something like intuition of iteration was perhaps held by Weyl or Poincaré, but it is difficult to get a very clear understanding of their views since their remarks on intuition are sparse. Brouwer's view seems to be that iteration and the structure of the natural numbers arise through the "self-unfolding" of the intuition, but not that either is an *object* of intuition. There are, however, difficulties in understanding just what Brouwer had in mind, for he speaks not only of constructing objects by such a process, but also of reflecting on our constructions. It is difficult to determine whether mathematical intuition is supposed to involve simply the process of constructing objects or whether it also involves *reflection* on our constructions. Brouwer clearly appeals to such reflection in intuitionistic developments in mathematics, for example, in his famous "proof" of the bar theorem, but whether we should think of the knowledge we have by reflection as intuitive is difficult to say.

8. SOME "EXTENDED" PROOF-THEORETIC VIEWS

Some of the literature in proof theory after the appearance of Gödel's incompleteness theorems has also suggested that we have intuitive abilities in mathematics that extend beyond what was encompassed by Hilbert's original conception. It is well-known that Gödel, following Bernays, concluded from his incompleteness results that we must "distinguish two component parts in the concept of finitary mathematics, namely: first the constructivistic element which consists in admitting reference to mathematical objects or facts only in the sense that they can be exhibited, or obtained by construction or proof; second, the specifically finitistic element which requires in addition that the objects and facts considered should be given in concrete mathematical intuition".[14] The objects of concrete mathematical intuition are said to be "finite space-time configurations of elements whose nature is irrelevant except for equality or difference". Gödel then points out that it is the second requirement that must be dropped if we are to give a consistency proof even for elementary number theory. The resulting epistemological situation is somewhat unclear. Gödel's approach in his *Dialectica* paper was to introduce "abstract" objects of a type not found in PA: primitive recursive functionals of finite type. But no epistemological account of these (abstract) objects is offered beyond saying that their introduction complies generally with a constructivist outlook. Gödel suggests in another place that giving a construc-

tive consistency proof for classical mathematics means "to replace its axioms (those of classical mathematics) about abstract entities of an objective Platonic realm by insights about the given operations of our mind".[15]

A number of logicians have suggested that there is a kind of intuitive evidence corresponding to the constructive but non-finitist parts of mathematics. The views of intuitionism might be taken as support for this. It is well-known that Gentzen also appealed to what he called "visualization" of steps in transfinite induction up to ε_0 in his consistency proof for elementary number theory. Notions similar to that of visualization have appeared in many other places in the literature as well. Kreisel, for example, developed this view to some extent in his picture of the "modified" Hilbert program. In fact Kreisel suggests that ε_0 is the exact limit of "idealized concrete intuition" in the sense that one can get to any ordinal $< \varepsilon_0$.[16] It should also be noted that the idea of visualizing a sequence of ω copies of a structure that we can visualize, taken together with reflection on the concept of ω-iteration, and the idea that we can reflect on the process of reflection, can be viewed as the underlying epistemological framework in the development of certain autonomous transfinite progressions of theories due to Kreisel and Feferman. It is fairly clear that Kreisel thinks of visualization as giving a kind of intuitive evidence, but the processes of reflection involved in building up a progression of theories presumably do not provide intuitive evidence. The role these processes would have in a wider philosophical account of mathematical knowledge is quite unclear. In fact it seems fair to say that from the philosophical side views about intuitive evidence in constructive mathematics that go beyond Hilbert's original conception of intuitive evidence have hardly been investigated at all.

9. GÖDEL ON SETS

In our brief survey of some main conceptions of mathematical intuition we should note how mathematical intuition has also been invoked in the context of set theory where it would be very difficult to think of the set theory involved as constructive. The best known views about this are Gödel's. In "What is Cantor's Continuum Problem?" and "Russell's Mathematical Logic" Gödel claims that we have something like an intuition of the objects of transfinite set theory. It is interesting that Gödel actually thought phenomenological considerations would be important in helping to determine which principles should be acceptable in setting up axioms of set theory and, in particular, axioms that would decide the continuum problem.[17] In speaking

of intuition in set theory what Gödel had in mind was transfinite set theory where sets are thought of as forming a cumulative hierarchy which is built up in stages by a process of iteration of the operation of "set of" applied to some given objects. I shall not describe the hierarchy here or the process by which it is obtained, but such a description will figure into my discussion later of HF sets. The HF sets are of course the sets obtained at the very earliest stages of the iterative process of forming sets that Gödel had in mind.

The point to note about Gödel's view is that he is willing to claim that the assumption that the transfinite sets obtained by this process exist is as legitimate as the assumption that there are physical bodies, and that there is just as much reason to believe in their existence. Gödel is thus claiming that there is an analogy in our evidence for these objects. At the same time he points out that such sets "clearly do not belong to the physical world and even their indirect connection with physical experience is very loose...". In a much quoted passage he says

> But despite their remoteness from sense experience, we do have something like a perception also of the objects of set theory, as is seen from the fact that the axioms force themselves upon us as being true. I don't see any reason why we should have less confidence in this kind of perception, i.e., in mathematical intuition, than in sense perception, which induces us to build up physical theories and to expect that future sense perceptions will agree with them and, moreover, to believe that a question not decidable now has meaning and may be decided in the future ... That new mathematical intuitions leading to a decision of such problems as Cantor's continuum hypothesis are perfectly possible was pointed out earlier.[18]

Gödel thus suggests that mathematical and perceptual intuition have analogous roles in our knowledge even in the case where the objects of mathematical intuition are the kinds of sets that can be shown to exist in transfinite set theory.

Gödel makes several further remarks about set-theoretic intuition meant in part to dispel what would seem to be obvious disanalogies between mathematical and perceptual intuition:

> It should be noted that mathematical intuition need not be conceived of as a faculty giving an *immediate* knowledge of the objects concerned. Rather it seems that, as in the case of physical experience, we *form* our ideas also of those objects on the basis of something else which *is* immediately given. Only this is something else here is *not*, or not primarily, the

sensations. That something besides the sensations actually is immediately given follows (independently of mathematics) from the fact that even our ideas referring to physical objects contain constituents qualitatively different from sensations or mere combinations of sensations, e.g., the idea of the object itself, whereas, on the other hand, by our thinking we cannot create any qualitatively new elements, but only combine and reproduce those that are given. Evidently the "given" underlying mathematics is closely related to the abstract elements contained in our empirical ideas.

In a footnote to this last sentence Gödel says that there is a close relationship between the concept of set explained in his paper and Kant's categories of pure understanding. The function of both is "synthesis", by which he means the generating of unities out manifolds. In Kant's philosophy this corresponds to the idea of being conscious of *one* object on the basis of its various aspects. Gödel continues by saying that it does not follow that the data involved in mathematical intuition, because they cannot be associated with actions of certain things upon our sense organs, are purely subjective, as Kant thought. "Rather they, too, may represent an aspect of objective reality, but, as opposed to the sensations, their presence in us may be due to another kind of relationship between ourselves and reality".

Some of these ideas are very difficult to understand. Further remarks of Gödel are reported in Wang's writings and Wang attempts to develop in more detail a view like Gödel's. One of the most interesting features of Wang's development is a kind of assimilation of notions of transfinite set theory to notions from constructive mathematics. I shall have more to say about some of Wang's ideas in my discussion of intuition and HF sets. A recent attempt to directly interpret the above passages has been made by Maddy. I shall briefly consider her interpretation of set-theoretic intuition in Chapter 7.

10. PLATONISM AND CONSTRUCTIVISM

Because Gödel is speaking of such large sets his views on set-theoretic intuition raise in a way that cannot be avoided the question of platonism or realism in mathematics. As it concerns the notion of mathematical intuition we can think of this as the question of whether the objects of intuition are mind-independent so that we have a universe of objects that we come to learn about or, rather, whether they are our own mental constructions so that we cannot suppose that they exist independently of these constructions. On the platonist thesis, statements about objects are true or false independently of

our knowledge or our constructions. My approach to this issue will be to develop some suggestions in Husserl's phenomenology according to which we could think of platonism and constructivism as compatible. Of course I shall not be examining the case of transfinite sets but I think the approach to the issue I discuss would have significance for these objects just as it will for natural numbers and finite sets. I shall be speaking of constructions and presenting what might be considered a constructivist view but I shall argue that, properly understood, this does not necessarily preclude platonism with respect to the existence of mathematical objects. One might even go farther and claim that Husserl himself actually adopted a platonistic position. This is a large issue in Husserl's philosophy, depending as it does on reconciling what might be called Husserl's epistemological idealism with his ontological realism. I shall offer some comments about this without attempting to make a case for any particular position on the issue.

11. MATHEMATICAL TRUTH AND MATHEMATICAL KNOWLEDGE

Questions that can be raised about platonism and constructivism should remind us of another large issue that lies in the background of any discussion of mathematical intuition: the kind of problem posed by Benacerraf about how to square an account of mathematical knowledge with an account of mathematical truth.[19] For as Benacerraf suggests, our best account of mathematical truth – Tarski's – depicts truth conditions in terms of conditions on objects like numbers and sets whose abstract nature places them beyond the reach of the better understood means of human cognition, like sense perception, so that it is unclear how we could have any mathematical knowledge whatsoever. In particular, numbers and sets, as conceived on such an account, would not be objects with which we could stand in causal relations, but causal accounts of knowledge, as Benacerraf sees it, are our best hope in epistemology. On the other hand, accounts that do better with mathematical knowledge do so at the expense of providing an account of mathematical truth. They fail to connect the conditions according to which we *know* that mathematical statements hold with conditions of the *truth* of such statements. They simply reinterpret mathematical language by offering some "non-standard" semantics – modal, intuitionistic, substitutional, or the like – which may characterize what it means for statements to be, for example, verifiable or provable, but not what it means for them to be true. The only way to escape this conclusion is to consider and actually reject the "standard" Tarskian account of truth, and to provide an alternative semantics

as a substitute. No one would presumably be prepared to actually reject a Tarskian theory of truth.

There have of course been many kinds of responses to this problem. One or more of the assumptions that permit framing of the problem may be challenged. Some writers wish to retain in some way the "standard" account of truth but to nonetheless make out an account of mathematical knowledge. Maddy has given a response of this sort specifically in relation to Gödel's ideas about set-theoretic intuition.[20] She wishes to show that causal theories of knowledge are reconcilable with set-theoretic realism. Another general kind of response to the problem is to challenge the idea that some kind of "non-standard" semantics cannot be taken to be a theory of truth for mathematical language. Responses of this kind have been developed by several writers. Michael Jubien, for example, has called into question the assumption that there is an adequate platonistic theory of truth, that is, an account of mathematical truth that "flows smoothly from the postulation of pure abstract entities".[21] He suggests how a non-platonistic account might be developed on the basis of modal notions. Another approach along these lines is suggested by Gottlieb.[22] Gottlieb argues that Benacerraf's problem is avoided by an interpretation of the language of mathematics as part of a language with substitutional quantification. On this interpretation, he argues, we incur no commitment to mathematical objects and so avoid epistemological problems with the "platonist" view of truth. Although Tarskian semantics is not retained, he believes that interpreting mathematical language "as an upper level of a language with substitutional quantification cannot be attacked as a failure to provide a theory of truth". In spite of what Gottlieb says by way of support for this remark there seem to me to be several serious difficulties with his approach. Some of the wider philosophical questions concerning anti-realism, "non-standard" semantics and the language of mathematics have of course been discussed by a number of philosophers, notably Dummett and Putnam.

In the course of this study I shall discuss some implications of a phenomenological view for issues about anti-realism, truth and knowledge. A number of Husserl's remarks will clearly have application in mathematics as will be seen already in the next chapter.

12. PRINCIPAL OBJECTIONS TO MATHEMATICAL INTUITION

The principal objections to the notions of mathematical intuition that I wish to consider in this study emerge clearly from writings on the subject. I shall

concentrate on two types of objections. From the point of view of a phenomenological account of mathematical intuition these objections will be closely related. First, there is the objection that mathematical intuition could not be a source of knowledge about objects analogous to perceptual intuition, which may be viewed as a source of knowledge about ordinary perceptual objects, because typical mathematical objects like natural numbers, sets and functions cannot, by virtue of their abstract nature, be ostended, could not be causally related to our senses, could not be objects in space-time, and are not individually identifiable and determinate. These last charges in particular would also be direct consequences of the view that mathematics is really not concerned with individual mathematical objects at all but rather only with certain *structures*. The structuralist view will be considered in more detail in Chapter 4.

I shall develop a response to this type of objection on which it is possible to say that there is something like mathematical intuition which is analogous to perceptual intuition in some important respects, but which at the same time recognizes some obvious specific differences between the two modes of knowledge. I shall want to develop the analogy without conflating mathematical and empirical knowledge, which appear to be different in some important ways. One of the differences of which I shall later take note has to do with the a priori character of mathematical knowledge.

The second objection is that even elementary mathematical knowledge, like arithmetical knowledge, could not be *founded* on intuition. If we consider natural numbers, for example, it is sometimes claimed that natural numbers may be represented in intuition by concrete symbol configurations, like finite strings of strokes, or in some other way that is available to ordinary sense perception. Our intuition of numbers is then supposed to be "founded" on these representations. So it would follow that our knowledge of small numbers and of singular statements about them would be provided by a kind of mathematical intuition founded on sense perception. But it is objected that on the basis of such representations we could not hope to derive any genuine *mathematical* knowledge. For our knowledge of signs and their properties would not be the same as knowledge of numbers and their properties. A related objection is that the perceptible representations are far from being sufficient to uniquely determine the mathematical objects involved. Even if we could offer an account of how our arithmetical knowledge is "founded" on perceptual intuition in these cases there would still be problems about how we could give any intuitive justification for claims about the existence of large numbers, or for *universal* arithmetical statements like

$(\forall x)(\forall y)(x + y = y + x)$ which are proved by mathematical induction. It is often argued that the kind of insight about the progression of intuitions needed to see that induction and related principles hold must be of a rather different character from the intuition of individual numbers, so that we could not be said to have this knowledge by way of intuition. At the very least, no account has been forthcoming as to how we could be said to have it by intuitive means. Attempts to account for such knowledge which take it to be founded on perceptible representations of numbers are circular: they can typically be shown to presuppose the very knowledge that intuition is supposed to make possible. I shall return to these objections in Chapter 4.

CHAPTER 2

THE PHENOMENOLOGICAL VIEW OF INTUITION

1. INTRODUCTION

In this chapter I shall present and discuss some of the most important material from Husserl's texts on intentionality and intuition, and the role of intuition in mathematical knowledge. The writings of Husserl show very clearly that on his conception of intuition there is an analogy between perceptual and mathematical intuition. A more detailed analysis of Husserl's views on ordinary perceptual intuition will follow in the next chapter by way of exploring and developing this analogy in later chapters. Some of Husserl's general remarks on the concept of truth and its relation to evidence (Evidenz) given by intuition will also be considered in this chapter, as will ideas about adequate, apodictic, a priori and intersubjective evidence. Especially relevant for the notion of mathematical intuition are Husserl's remarks on founding, acts of abstraction, acts of reflection, and the idea of hierarchies of acts built up by reflection and abstraction. Some of the most important texts on these matters will be considered.

In a number of his works Husserl develops what is in effect a unified account of intuition *of* and intuition *that* which should have applications in different domains of knowledge. Husserl does not get very far with anything that could be considered an account of mathematical intuition. Later writers like Weyl, Becker, Heyting and possibly Gödel could be viewed as having carried some Husserlian ideas farther in mathematics. Husserl's own analyses and illustrations are almost always concerned with the phenomenology of perceptual intuition of ordinary physical objects. In his earliest writings on arithmetic he does speak of the fact that numbers can be intuitively present in our experience but many of the details of his view of such intuition are only developed later. In particular, discussion of the "categorial activities" involved in "categorial intuition" of numbers and other abstract objects is first undertaken in the context of a mature theory of intentionality in the Sixth Investigation of the *Logical Investigations*. Husserl takes up similar questions again in Part II of *Formal and Transcendental Logic* and in portions of *Experience and Judgment*. In these works he gives an account of intuition

(Anschauung) and evidence (Evidenz) suitable to mathematical objects, but many of the main ideas are obscure and only partially worked out. His observations on the specific question of what constitutes evidence, or what it means to "intuit" or have an analog of a "representing content" for objects like natural numbers or finite sets leave much to be desired since they are worked out in the general setting of his theory of intentionality without regard for issues or technicalities concerning these objects in particular. Of course one might still expect to find some of the most important processes involved in the intuition of numbers and finite sets already analyzed to some extent in *Philosophie der Arithmetik*. In fact, I shall claim as much. But there are still many difficulties, especially in light of many issues about numbers and sets that have emerged since Husserl's time. It is especially difficult to see how an account of intuition and evidence should be developed in the case of mathematical objects about which Husserl has very little to say, e.g., Cantorian transfinite sets.

In spite of problems with understanding details of Husserl's view of intuition in mathematics several themes do clearly emerge from his work.

2. INTENTIONALITY AND INTUITION

To begin, it must be realized that Husserl's views on intuition are worked out in the context of a general theory of intentionality.[1] Unlike the views on mathematical intuition mentioned in the last chapter – with the exception of Kant – the phenomenological view begins with a substantial philosophical theory of mental acts and processes in which the notion of intuition has a distinct role. Mathematical beliefs, like many kinds of cognitive acts, have the characteristic of being "about" something, of being directed toward objects. Mathematical beliefs are typically beliefs *about* numbers, sets, functions, spaces, groups, and so on. Mathematical beliefs are often thought to be more difficult to analyze than ordinary beliefs about the physical world because the "objects" of mathematical beliefs are evidently quite different from ordinary physical objects. Phenomenologists, however, could argue that the purported differences between mathematical and physical beliefs are not so great as they are often made out to be. For in the case of beliefs about either mathematical or physical objects it is the "intentional object" that we are to consider. In practice this means that we are to consider the cognitive *acts* in which objects are intended, and how the objects are intended in those acts, without making naive metaphysical assumptions about the objects. It should perhaps be noted that that the consideration of cognitive acts is not to

be taken to require commitment to introspective reports, in the style of introspectionist psychology. Rather, it depends, in effect, on transcendental arguments about the possibility of experience, and in this respect it is on the same footing as much of the recent work in cognitive science. This is a point that should be significant for philosophical questions about mathematical cognition, especially in intuitionistic mathematics, since there is need for clarification about the intuitionistic view that we have access to the mental constructions that make up the subject matter of mathematics.

A standard way to view the concept of intentionality is to say that acts of cognition are directed toward objects by way of the "content" of each act, where the object of the act may or may not exist. We might picture the general structure of intentionality in the following way,

Act (Content) --------> [object],

where we "bracket" the object because we do not assume that the object of an act always exists. Phenomenologists are famous for suggesting that we "bracket" the object, and that we then focus our attention on act (noesis) and act-content (or noema), where we think of an act as directed toward a particular object by way of its content (or noema). Whether the object exists or not depends on whether we have evidence for its existence, and such evidence would be given in further acts carried out through time.

We can capture what is essential (for our purposes) to the distinction between act and content by considering the following cases: a mathematician M might believe that S, or might know that S, or might remember that S, where S is some mathematical statement. In these cases different types of cognitive acts are involved believing, knowing, remembering but they have the same content, expressed by S. The act-character changes, but the content is the same. Of course the content may also vary while the act-character remains the same. The content itself can have a structure that is quite elaborate. Also, when we say that the content is "expressed" by S we shall mean that the mathematical *proposition* S is an expression of the content of a particular cognitive act. Thus, there is a direct parallel between intensionality, a feature of language or expression, and intentionality, a feature of cognition, insofar as we are restricting attention to those expressions which are expressions of cognitive acts. We should not necessarily expect, for example, substitutivity *salva veritate* and existential generalization to hold for inferences involving expressions of intentions. In most of this work we shall regiment our understanding of what forms S can take in terms of the syntax of first-order theories. Thus when we say that an act is directed toward an

object we can express this with the usual devices of first-order theories: individual constants and bound variables. For example, a mathematician might believe that Fa for a particular F and a particular a, or that $(\exists x)Fx$ for a particular F and a particular domain of objects D. The restriction to first-order theories is not necessary, it simply makes for a tidy beginning.

Intuition is understood in terms of the notion of *fulfillment* of intentions. Intentions may be thought of as fulfilled, partially fulfilled, frustrated, and so on. We *intuit* an object when our intention to the object is fulfilled, for then the object intended is actually displayed or presented, precisely *as* it is intended, and this is not the case where the object intended is not intuited.[2] The difference between intuiting an object and merely having an intention directed toward the object can be roughly described as the difference between actually seeing something and merely thinking about it or having a concept of it. There is an obvious resemblance here to Kant's distinction between having a concept of an object and having an intuition of the object, even though as we shall see later Husserl believed that he was extending and developing a Kantian view of knowledge for areas of human experience – logic and mathematics – that Kant did not fully appreciate. In the case of mathematical objects like natural numbers and finite sets the problem therefore is to give an account of just what is to count as fulfillment of mental intentions to such objects.

In Husserl's philosophy intuition is typically to be viewed as a process in which certain acts must be carried out in order for an intention to be fulfilled. It is necessary to have certain *experiences*. Husserl's views on intentionality have of course been written about at length and I shall certainly not attempt to do justice to them here.[3] An important theme for the topic under discussion, however, is that intentions can be thought of as referring not only to objects that might exist independently of us or of our constructions, but also to certain kinds of *expectations* or *anticipations* about the course of our experience with respect to such objects. An "empty" intention is one which from this point of view is a pure expectation or anticipation about an object. At a given point in our experience of an object the expectation may be more or less well-determined, but there will always be at least some expectation about the object depending upon how it is intended at that point. Associated with every intentional act will be a *horizon* of possible experience in which the intention to an object would come to be fulfilled, partially fulfilled, frustrated, and so on. The horizon associated with an intention indicates a more or less well-determined procedure or process for finding the object emptily intended, or for making our knowledge of it more determinate. In the

best circumstances the process will be quite well-determined. We might characterize it, for example, by a certain set of functions. In certain domains of knowledge, especially in mathematics, the "functions" could perhaps be specified very precisely, e.g., different classes of recursive functions. Husserl actually discusses the role of algorithms in arithmetic to some extent in unpublished writings. For our purposes the general notion of "horizon" will prove to be important, and it will be discussed in more detail in the following chapter.

Intuiting an object amounts to actually carrying out the acts specified by the process or procedure in which the objects or objects would be given. In that case we actually see what was merely anticipated or thought about. It is clear from Husserl's writings that what we should regard as *possible* in intuition or experience is to be determined on the basis of what a mathematician M can *actually* intuit or experience. The modal notions here are to be analyzed with reference to M. M could not hope, for example, to carry out a sequence of perceptions of an object if for some reason she could not intuit a part of the object, even though she may have a conception of such a sequence and of a being who would carry it out. In the most developed part of Husserl's views the idea of what M can intuit or experience would have to be understood in terms of the "transcendental ego". An analysis of the transcendental ego would show what types of experience are possible for the kind of beings that we are. It seems clear at any rate that we should not suppose that every kind of experience of which we can conceive is something that we could possibly carry out. There may be, for example, certain well-formed mathematical statements that express intentions (and hence are meaningful) that could not possibly be fulfilled or even partially fulfilled by beings like ourselves. In that case we could not regard them as expressions of knowledge of objects, even though they may have some meaning for us. In other cases it may be that a statement is not in principle incapable of fulfillment, but only that certain contingencies prevent it from being fulfilled. The analysis of M – ultimately the transcendental ego for Husserl – is thus integral to an account of mathematical knowledge. Some have even thought we should be committed to treating it as a part of mathematics.

3. INTUITION OF ABSTRACT OBJECTS

It is very clear that Husserl regards knowledge to be a product of (empty) intentions and intuition. There is no knowledge without intuition. In the *Ideas*, for example, Husserl states as the "principle of all principles" that

"every primordial dator intuition is a source of authority (Rechtsquelle) *for knowledge*, that *whatever presents itself in "intuition" in primordial form* (as it were in its bodily reality) *is simply to be accepted as it gives itself out to be...*".[4] The fourth section of *Ideas*, along with the Sixth *Logical Investigation*, Part II of *Formal and Transcendental Logic*, *Cartesian Meditations*, and various other writings make it very clear that intuition is a condition for knowledge quite generally, not only for knowledge that we have by sense-perception. Husserl first speaks of a "widened", "non-sensuous" concept of intuition in the *LI*. The objects of this kind of intuition are "categorial objects" of which Husserl gives as examples "aggregates, indefinite pluralities, totalities, numbers, disjunctions, predicates, states-of-affairs".[5] These objects can be given in acts that must count as intuitive in a sense analogous to that of acts of ordinary perceptual intuition. In his initial discussion of a widened concept of intuition in *LI* Husserl says that

> there must at least be an act which renders identical services to the categorial elements of meaning that merely sensuous perception renders to the material elements. The essential homogeneity of the function of fulfillment ... obliges us to give the name 'perception' to each fulfilling act of confirmatory self-presentation, to each fulfilling act whatever the name of an 'intuition', and to its intentional correlate the name of *'object'*. If we are asked what it means to say that *categorially structured meanings* find fulfillment, confirm themselves in perception, we can but reply: it means only that they relate to the object itself *in its categorial structure*. The object with these categorial forms is not merely referred to, as in the case where meanings function purely symbolically, but it is set before our very eyes in just these forms. In other words: it is not merely thought of, but intuited or perceived ... What shall we call the correlate of a non-sensuous subject-presentation, one involving non-sensuous structure, if the word 'object' is not available to us? How shall we speak of its actual giveness, or apparent giveness, when the word 'perception' is denied us?[6]

Husserl concludes that we must therefore draw a quite general distinction between sensuous and categorial intuition, or at least show the possibility of such a distinction.

Following this distinction, sensuous objects are characterized as objects at the "lowest" possible level of intuition, while categorial objects are said to be objects at "higher levels". Husserl spends a good deal of time working out general features of this characterization. A notion of "founding" is introduced in the Sixth Investigation according to which acts in which categorial objects

are given are said to be founded on underlying acts, ultimately on acts of ordinary sense-perception. We should take note of some of the texts concerning this notion now, reserving until later the attempt to develop and elaborate on the notion in the context of number theory and set theory.

In ordinary sense-perception an object is said to be apprehended in a "straightforward" (schlicter) manner: it is an immediately given object which is not constituted in acts founded on other acts:

> Sensuous objects are present in perception *at a single act-level*: they do not need to be constituted in many-rayed fashion in acts of higher level, whose objects are set up for them by way of other objects, already constituted in other acts.[7]

Husserl goes on to say that each straightforward act of perception, by itself or with other acts, can serve as a basic act for new acts which bring a *new* awareness of objects which essentially presupposes the straightforward act(s). The following passage is especially important for it also tells us that the manner of appearance of the new objects is essentially determined by the founding acts:

> What we have are acts which ... *set up new objects*, acts in which something *appears as actual and self-given*, which was not given, and could not have been given, as what it now appears to be, in these foundational acts alone. *On the other hand, the new objects are based on the older ones, they are related to what appears in the basic acts.* Their manner of appearance is essentially determined by this relation. We are here dealing with a sphere of objects, *which can only show themselves 'in person' in such founded acts*. In such founded acts we have the categorial element in intuition and knowledge, in them assertive thought, functioning expressively, finds fulfillment; the possibility of complete accord with such acts determines the truth, the rightness, of an assertion.[8]

In ordinary sense-perception an individual object appears "all at once", as soon as our glance falls upon it. No apparatus of founding and founded acts is required in order for the individual to appear. Husserl thus says the individual given in straightforward perception is present in an *immediate* way, while categorial objects are given "mediately" via acts that must first be carried out in order to constitute a "higher-order" object. Continuing his contrast of ordinary perception with categorial intuition Husserl says that

> The unity of ordinary perception does *not* therefore arise through *our own synthetic activity*, as if only a form of synthesis, operating by way of founded acts, could give unity of objective reference to part-intentions. It requires no articulation and hence no actual linkage. The unity of perception comes into being as a *straightforward* unity, *as an immediate fusion of part-intentions, without the addition of new act-intentions.*[9]

Some form of synthesis is involved in either ordinary perception or in mathematical intuition. In the latter case an *active* synthesis is involved, while in the former case it is a form of *passive* synthesis.[10] In the case of the active synthesis involved in mathematical intuition we must actually undertake some acts in order to obtain the object. We must, in effect, "construct" the object. In the case of passive synthesis the object is immediately given, we do not need to actively engage in any particular acts in order to see it. In ordinary perceptual intuition no new type of act over and above the acts of continuing the perception of the object would need to be undertaken in order to be aware of the perceptual objects. Such a new act would have a different "thetic character". The notion of "thetic character" is a technical term in the more mature analysis of noemata in the *Ideas* which has its roots in the notion of the act "quality" in *LI*. In Husserl's analysis of the noetic, noematic, and hyletic aspects of consciousness in the *Ideas* several components of the noetic/noematic structure are characterized: thetic character, the Sinn component, and the "nucleus" of the Sinn component which Husserl calls the "determinable x". Thetic character simply corresponds to the type of act or attitude, e.g., perceiving, remembering, imagining, and so on. In founded acts of the type of which Husserl is speaking in the above passage the character of the act would not be that of "perceiving", but would have to be of the variety appropriate to mathematical intuition.

In order to further clarify his point Husserl says that we might of course attempt to improve our knowledge of an ordinary perceptual object so that we do undertake further acts to see the object from different perspectives, to handle it, smell it, and so on. But in this process there is no question of "founding" in the sense relevant to the intuition of categorial objects, for the further acts involved will continue to have the same thetic character as the earlier acts – that of perceiving the object. Husserl points out that perception of the individual as a whole can be said to be founded on perception of its parts, but that this is a different sense of "founding", a sense relevant to the Third Investigation. For categorial objects a new type of act would be

required which would be grounded in the acts that underlie it and would be unthinkable apart from these. It should be noted that both types of founding will be involved in mathematical intuition. The intuition of an object as a whole will be founded on the intuition of its parts, but acts of mathematical intuition will also be founded on underlying acts of perception. In Husserl's later writings it becomes clear that the acts involved in the intuition of objects like natural numbers and finite sets are typically acts of collecting, reflecting, abstracting, and comparing. Acts of imagination may also play a role.

After further analyzing the distinction between sensuous and categorial intuition in *LI*, Husserl concludes that "It is *essential* to these (founded) acts,

> in which all that is intellectual is constituted, that they should be achieved in stages. Objectivations arise on a basis of objectivations, and constitute objects which, as objects of higher order, objects in a wider, intellectual sense, can only come to light in such founded acts. This excludes from synthetic acts that immediate unity of representation which unites all representative contents of straightforward intuitions. The complete synthetic intuition therefore arises ... in so far as the mental content which binds the underlying acts itself sustains *interpretation* as the objective unity of the founded objects ....[11]

This passage is significant for its explicit mention of the fact that founded acts are built up in stages from what is already given. This kind of idea has of course played a role in certain conceptions of mathematical objects, especially in constructivist conceptions. Toward the middle of the passage Husserl begins to speak of an analog in acts of categorial intuition of the "representing content" in straightforward intuition. The representing content in straightforward intuition is the component of perception contributed by sensation. Husserl sometimes refers to sensory data as the "matter" or the qualitative component which sustains a form of interpretation in an act of perception. The same sensory data may sustain quite different interpretations, as in the case of Gestalt shifts, but they still function to determine and constrain what the perception could be about. That there must be something analogous to representing content in the intuition of abstract object cannot be denied according to Husserl. For the act structure in the two cases is supposed to be fundamentally the same. Representing contents, he says in section 55 of *LI*, constitute the difference between empty and fulfilled intentions – they are responsible for fulfillment: "Only intuitive acts render their object apparent, seeable, for the reason, namely, that a representing content is there, which the interpretative form sees as a likeness, or as the

very self, of the object."

In the *LI* Husserl admits that there are difficulties in trying to understand what the representing content could be in the case of abstract objects. The function analogous to representing content in the intuition of abstract objects could not be performed by what is given in the underlying, founding acts because the founded acts are supposed to have a new object. In the passage at hand, Husserl therefore says that "the mental content which binds the underlying acts 'must itself' sustain interpretation as the objective unity of the founded objects."

What Husserl evidently has in mind here is that categorial intuition must typically be carried out in stages, or in a sequence of acts. We do not have an immediate unity of representation as in straightforward intuition. Hence Husserl speaks of it as "synthetic" in a way that straightforward intuition is not. In categorial intuition the "unity of representation" must be synthetically constituted. A categorial intuition would evidently be "complete" when this process had been carried out and the result looked at as a new object. My interpretation of Husserl's somewhat obscure remark that "the mental content which binds the underlying acts" must itself "sustain interpretation as the objective unity of the founded objects" will be developed along these lines. Acts of reflection and abstraction make it possible to regard the result of the process as a new object. Examples of the relevant kind are suggested by analyses of the process of "collecting" given in *PA*. In "collecting" we carry out a certain process and then regard the result of this process as a new object, a collection. In this process we evidently have the analog of representing content since the underlying process we carry out will fix what it is that we can regard as a new object. It will be an analog of representing content for categorial intuition insofar as only the *structural* features of the underlying process are relevant. What should be meant by this last remark is something that will be discussed in detail in Chapters 5, 6, and 7. I think the matter of the analog of representing content for acts of categorial intuition would be extremely difficult if not impossible to understand solely on the basis of Husserl's remarks in *LI* but, as has been indicated, I shall attempt some elucidation of it in later chapters using ideas from constructive mathematics.

Of course sensory data would not be involved in a direct way in the analog of representing content for acts of categorial intuition, while such data would be directly involved in ordinary perceptual intuition. What is relevant about the underlying process in categorial intuition is its structural features, not the qualitative features associated with the acts involved. Husserl nonetheless says that categorial acts must be ultimately founded on, or have their

"origins" in, acts of sense-perception. Any "regress of founding" has to eventually come to an end. The structure of this process will be of special interest in our later considerations. For now we might note that there seems to be a definite resemblance here to the older Kantian theme that while all knowledge begins from sense experience it is not limited only to sense experience. In a passage that has this Kantian flavor Husserl thus says

> It lies in the nature of the case that everything categorial ultimately rests upon sensuous intuition, that a 'categorial intuition', an intellectual insight, a case of thought in the highest sense, without any foundation of sense, is a piece of nonsense. *The idea of pure intellect*, interpreted as a faculty of pure thinking (=categorial action), *quite cut off* from a 'faculty of sensibility', could only be conceived *before* there had been an elementary analysis of knowledge in the irrefragable evidence of its being.[12]

This view is repeated throughout Husserl's writings and occurs in essentially the same form even in *Experience and Judgment*. The idea that founded acts should ultimately be founded on acts of sense-perception is supposed to have significant epistemological consequences on Husserl's view, some of which we take up in the sequel.

4. ACTS OF ABSTRACTION AND ABSTRACT OBJECTS

An important process involved in categorial intuition is that of categorial or formal abstraction. Husserl says that

> If we ponder on the peculiarity of eidetic abstraction, that it necessarily rests on individual intuition, but does not for that reason mean what is individual in such intuition, if we pay heed to the fact that it is really a new way of conceiving, constitutive of generality instead of individuality – *then the possibility of universal intuitions arises, intuitions which not merely exclude all individuality but also all sensibility from their intentional purview*. In other words, we distinguish between *sensuous abstraction*, which yields *sensuous concepts* – purely sensuous or mixed with categorial forms – and *purely categorial abstraction*, which yields *purely categorial concepts* ... Sensuous concepts find their immediate basis in the data of sensuous intuition, categorial concepts in the data of categorial intuition, purely with regard to the categorial form of the whole categorially formed object.[13]

Now the view of abstraction that Husserl has in mind here, and the many

difficulties associated with abstraction are discussed in some detail in the Second Logical Investigation. There are different kinds of abstraction and we shall be primarily concerned with formal or categorial abstraction. Husserl, unfortunately, does not have much to say about what is involved in carrying out acts of abstraction necessary for the awareness of specific abstract objects. We can nonetheless make a number of general observations about acts of formal abstraction that will be of use in our treatment of number theory and set theory.

The notion of abstraction has of course been in bad repute for some time. I shall not attempt to offer a theory of acts of abstraction here but I would want to defend the view that on the basis of a proper phenomenological understanding of abstraction it would be difficult to deny that there are such acts. My approach will be to simply recognize a certain phenomenon and to make a few elementary observations about it. The phenomenon is one on which I think we could all agree, namely, that we are able to discern the *form* or *structure* of given phenomena in many kinds of circumstances that are relevant to the more theoretical parts of our knowledge. One finds many examples of success at this kind of thing in logic and mathematics. In logic, for example, we have certain beliefs about the form of a given sentence, or about the structure of a given argument. It is just such structural features we attempt to express in developing an adequate formalism. One can point to the same phenomenon in mathematics. In fact, I think we would be warranted in making the Kantian observation that without acts of formal abstraction we would not have logic and mathematics as we know it. As we are understanding the matter, acts of formal abstraction are just what make logic and mathematics possible.

Now some general observations we can make about acts of formal abstraction follow from the fact that we must view such acts as intentional, just like any other acts. They have the same basic structure as any intentional acts and hence can be fulfilled, partially fulfilled, frustrated, and so on, in the process of trying to carry them out. Acts of abstraction are, moreover, founded acts. They depend on the fact that certain acts have already been performed or "lived through". We then abstract from what is given in the prior act or acts – we attempt to obtain the form or structure of what is given. This involves a shift in thetic character. For we start with a given phenomenon and then view it in a different way from how it was originally given. In the example from logic we might be given a particular sentence in ordinary language. We can then approach this sentence with the intention of determining its form, relative to our background interests. In attempting to do

this we would be "abstracting" from other features of the sentence. The result would be a formal or structural abstraction. In our attempt to determine the form we may fail or succeed, just as in the case of our attempts to fulfill other kinds of intentions. Because acts of formal abstraction can be characterized by the distinction between mere intention and intuition we must view them as processes that produce certain beliefs, or that give evidence for certain beliefs about the structure of what is given. These beliefs could be incorrect – the intentions could be frustrated. Hence we may have better or worse evidence for the beliefs produced by this process. Since the requirements of an account of number or set abstraction are, I think, quite specific to these objects I shall take up questions about acts of abstraction again when I come to these objects. Some remarks that Husserl makes already in *PA* about number abstraction will prove to be helpful.

For the interpretation of our later remarks we should note that in the Second Logical Investigation Husserl is especially concerned to set aside nominalistic and psychologistic views of abstraction of the type developed by the empiricists Locke, Berkeley, and Hume. While not wishing, for example, to psychologically "hypostatize" the abstract objects obtained as a result of acts of abstraction he also wants to avoid the traditional metaphysical hypostatization of abstract objects associated with some forms of platonism.

It is clear in the Second Logical Investigation that Husserl recognizes intuition of concrete, space-time objects, like the individuals one intuits in ordinary sense-perception, but he also recognizes that there is intuition of abstract objects which have concrete instances – what might be called quasi-concrete objects. There is of course also intuition of objects which do not have concrete, space-time instances. Husserl wishes to say that any intuition of abstract objects will be founded on or will have its origin in straightforward intuition, but it will not be the same thing as intuition of individuals in sense-perception. Husserl's principal argument for this claim is to point out that intentions directed to individuals in ordinary perception are obviously different from intentions directed to objects like natural numbers or finite sets. In an act directed toward some number it must be the number itself that is the object of the act, not anything else. For what we *mean* in the act is the number, and for such an act to be fulfilled it is precisely the object intended that must be given. It would not do, for example, to suppose that a string of symbols given to sense-perception could count as the object of such an act. Strings of symbols and their properties, whether we view them as tokens or types, are quite different from numbers and their properties, and knowledge of the former could not count as knowledge of the latter. Simple arguments

about substitutivity *salva veritate* in the context of epistemic operators should serve to make this Husserlian point clear.

It seems to me that one of the important points of this argument is that in phenomenology, as Husserl became fond of saying, we must begin with the phenomenon just as it presents itself. In mathematical experience the objects with which we are concerned are not given as concrete individuals of sense-perception. So we must begin with the experience as we actually have it, without seeking from the beginning to reduce it to something else. The nominalist is not interested in an analysis of mathematical experience in these terms – he has a different purpose. Now I do not think that in phenomenology we should necessarily feel compelled to reject nominalistic reductions of objects if they could be given. However, even if we consider some of the best recent work on this program by philosophers like Field and Boolos there are still substantial objections to supposing that such reductions can be carried out for numbers and finite sets, and in parts of mathematics that go beyond elementary number theory and finite set theory in their ontological commitments it is even clearer that such reductions will be very difficult to provide.[14]

Husserl's refutation of psychologism involving abstract objects is well known. I shall not consider his various arguments here but it should be clear that mathematical objects may not be construed as mental objects either. Their properties are of an altogether different type from the properties of mental objects and it would not do to suggest that intuition of the former can be construed as intuition of the latter. Husserl appeals to much the same type of argument that he uses to bypass nominalism, and the argument is bolstered by his extensive critique of psychologism. We should also note at this point that on a Husserlian view intuition of an object like a number is not to be thought of as intuition of a mental image of some kind, nor does it require accompaniment of a mental image. It is not as if I must have some special mental occurrence or some image in mind to intuit an object like a number. On the contrary, to intuit a number is simply to carry out some acts in which a determination of number is made. We obviously do this all the time and no special images are or need to be involved.

Because the objects of acts of mathematical thought cannot be construed nominalistically or psychologistically it does not follow that such acts are directed to fictional or nonsensical objects. Husserl says that

> It is naturally not our intention to put the *being of what is ideal* on a level with the *being-thought-of which characterizes the fictitious or the*

> *nonsensical.* The latter does not exist at all, and nothing can properly be predicated of it: if we none the less speak of it as having its own, 'merely intentional' mode of being, we see on reflection that this is an improper way of speaking. There are, in fact, merely certain necessary and valid connections among 'objectless ideas', whose analogy with truths governing ideas having objects, has prompted this talk of objects merely presented which do not genuinely exist. Ideal objects, on the other hand, exist genuinely. Evidently there is not merely a good sense in speaking of such objects (e.g. of the number 2, the quality of redness, of the principle of contradiction etc.) and in conceiving them as sustaining predicates: we also have insight into certain categorial truths that relate to such ideal objects. If these truths hold, everything presupposed as an object by their holding must have being. If I see the truth that 4 is an even number, that the predicate of my assertion actually pertains to the ideal object 4, then this object cannot be a mere fiction, a mere *facon de parler*, a mere nothing in reality.[15]

This passage is important for its emphasis on the distinction between "objectless ideas" among which there can nevertheless be "necessary and valid connections", and, on the other hand, ideas having objects. Husserl would no doubt claim that some intentions, suitably understood, could be shown to be objectless in principle, e.g., intentions about roundsquares, largest prime numbers, etc. Even though statements of this type may be provably objectless they can still be involved in various types of logical relationships. This will be quite important for some of our comments later about mathematical knowledge. For even in the case of these kinds of statements we may have purely "logical" knowledge regarding the statements involved, that is, knowledge about the logical relationships in which such statements may be involved. But this is different from having knowledge of objects. For the intentions expressed by such statements could not possibly be fulfilled. In the case of fictional objects the manner of fulfillment, if one can speak of it at all, would have to be quite diferent from fulfillment in the case of numbers and finite sets. One of the more obvious differences concerns the way we *use* fictions and the way we use numbers and finite sets. Most philosophers would probably argue that we do not *apply* fictions at all, or that, if we do, it is certainly in a way very diferent from applications of number. One might argue, as Frege has suggested, that applicability requires not only that statements have a sense or intension but also that they have a reference.

Especially interesting cases arise, from the point of view of mathematical knowledge, when we do not know whether statements are objectless or not. In these cases we might admit that logical relationships among such statements give us a kind of "hypothetical" or "conditional" knowledge about objects, and yet this is still certainly different from knowledge we have where intuition of objects is involved, where there are objects to which our statements refer and we have methods for obtaining these objects.

In his comments following the above passage Husserl goes on to say that the meaning of "being" and of predication for abstract objects does not coincide exactly with their sense in the case of objects of ordinary perceptual intuition. In fact, there is a "fundamental categorial split in our unified conception of being". We are simply taking account of this split when we distinguish between "ideal" and "real" being, or between being as species and being as what is individual. In *FTL* Husserl returns to this theme and claims that the kind of existence an object is thought to have is itself constituted in the processes involved in intuition of the object, from one's experience of it. The meaning of the being of abstract objects is said to be derived from or founded on the sense of the being of the "real" objects of ordinary sense-perception.

5. ACTS OF REFLECTION

Apart from acts of abstraction certain kinds of acts of reflection are supposed to also be involved in the intuition and hence knowledge of abstract objects. One of Husserl's most detailed later discussions of acts of reflection is to be found in sections 77-82 of *Ideas*. Here too Husserl's treatment is very general and as a result not very helpful for an account of the intuition of mathematical objects. It is clear, as in the case of acts of abstraction, that acts of reflection have the same basic structure as any other intentional acts. They are called acts of reflection because they are directed to other acts or mental processes. They must be, of course, founded acts, since they depend on the fact that certain other acts have already been performed or "lived through", and they then have these acts as their object(s). Presumably there exist acts directed to mental acts which lack objects, e.g., acts that give sense to empty definite descriptions of acts. In saying that acts of reflection are founded we shall mean that they are founded on the acts on which they reflect, where these acts are then their object. It would seem to follow that in the case of reflection we have something like direct reference in that the act depends for its existence on the existence of its object.

It is important for my purposes that acts of reflection are object-directed acts. For reflection is then to be viewed as a process which produces beliefs, or which gives evidence for certain beliefs, just as in the case of other kinds of acts. Certainly some beliefs that we have about acts or about the structure of mental processes are more reasonable than others, so that we have "evidence" for some of these beliefs. Acts of reflection are especially important because, as directed to mental processes, they have as their objects the material which can "sustain interpretation as the objective unity of the founded objects"; in other words, the material that plays the analog of representing content in the intuition of abstract objects. In the case of either ordinary perception or mathematical intuition the representing content is supposed to be the material that gets "interpreted" by an act which is directed toward a certain object. The representing content figures into the intuition of an object, constraining and determining the intuition, but of course it is not itself the object of the intuition. If we are interested in the conditions necessary for intuition of abstract objects then evidently acts of reflection provide the material which gets interpreted by acts of abstraction, yielding intuition of "abstract" objects. In Chapter 5 I shall describe this in some detail in connection with specific examples.

Reflection has an important role in our knowledge because just as in the case of other acts, acts of reflection can be fulfilled, partially fulfilled, and so on. Acts of reflection are characterized, that is, by the distinction between intuition and empty intention. Husserl suggests of course that we can reflect on our reflections so that we obtain a hierarchy of reflections with earlier acts of reflection providing a foundation for later acts of reflection. The reader familiar with Frege's hierarchy of senses and references might find it useful to think of Husserl's view here as something like the cognitive counterpart to Frege's conception. Just as in the case of acts of abstraction I shall discuss reflection in the specific context of natural numbers and finite sets where it is possible to be more definite about the nature of the acts involved.

We might note that our comments thus far suggest that "straightforward" acts, like those of ordinary perception, are prereflective acts. Reflective acts could evidently not be identified with prereflective acts under any conditions. A hierarchical structure is forced by the fact that we are to think of the act (or the meaning or intention of the act) reflected on as the *object* of the act of reflection. The meaning or intention associated with the reflective act would be different from that associated with the act reflected on. And as we have said, the former could not exist without the latter. The reflective act would have a new content. The circumstances of the occurrence of the two kinds of

acts would no doubt be different. In reflective acts one can first differentiate what is reflected on from other things, compare, note similarities, note species/genus relationships, and so on. These possibilities are implied by reflection, but not by the original prereflective awareness.

6. TYPES AND DEGREES OF EVIDENCE

In *FTL* Husserl returns to the theme of the analogy between the structure of the mental acts involved in perceptual and mathematical intuition by discussing the kind of evidence that we have in each case. He says that

> The evidence of irreal objects, objects that are ideal in the broadest sense, is, in its effect, quite analogous to the evidence of ordinary so-called internal and external experience, which alone – on no other grounds than prejudice – is commonly thought capable of effecting an original Objectivation. The identity and, therefore, the *objectivity* of something ideal can be directly "seen" (and, if we wished to give the word a suitably amplified sense, directly experienced) with the same originality as the identity of an object of experience in the usual sense.[16]

It is pointed out that in either case deception is possible simply because we cannot have "absolutely apodictic evidence". Evidence is not to be seen as "an absolute criterion of truth". It could not be since the acquisition of evidence is part of an ongoing process in which further acts would be involved in improving the evidence that we have. Now although evidence will be provided by either perceptual or mathematical intuition the type of evidence that we have in each case will be different because the objects are of different types:

> *Category of objectivity and category of evidence are perfect correlates. To every fundamental species of objectivities* – as intentional unities maintainable throughout an intentional synthesis and, ultimately, as unities belonging to a possible "experience" – *a fundamental species of "experience", of evidence, corresponds*, and likewise a fundamental species of intentionally indicated evidential style in the possible enhancement of the perfection of the having of an objectivity in itself.[17]

After discussing more details about evidence and about hierarchies of evidences founded on the evidence given in perceptual intuition, Husserl makes some very important observations on the relation of the concept of truth to evidence. He notes that in his earlier work he had not sufficiently

appreciated how a phenomenological conception of evidence would have to be developed. In *FTL* and *CM* the matter becomes very clear. Husserl suggests that truth is something like a Kantian regulative idea, "lying at infinity":

> What if each and *every* truth about reality, whether it be the everyday truth of practical life or the truth of even the most highly developed sciences conceivable, remains involved in *relativities* by virtue of its essence, and referable to *"regulative ideas"* as its norms?[18]

He then asks an intriguing question to which I shall return at the conclusion of this essay:

> What if the relativity of truth and of evidence of truth, on the one hand, and, on the other hand, the infinitely distant, ideal, absolute truth, beyond all relativity – what if each of these has its legitimacy and each demands the other?

I shall be interested in this question from the point of view of the contrast between mathematical truth and the mathematical evidence given by intuition, for in this context the question has, I believe, a bearing on how we should view the relationship of platonism to constructivism. Husserl summarizes his view on the relation of evidence to truth in a way that bears quoting in full:

> ... we have continuously anew the *living truth from the living source, which is our absolute life*, and from the self-examination turned toward that life, *in the constant spirit of self-responsibility*. We have the *truth* then, not as falsely absolutized, but rather, in each case, as within its *horizons* – which do not remain overlooked or veiled from sight, but are systematically explicated. We have it, that is to say, in a living intentionality (called 'evidence of it') whose content enables us to distinguish between 'actually self-given' and 'anticipated', or 'still in our grip' retentionally, or 'appresented as alien to the Ego's own', and the like – a content that, with the uncovering of the attendant intentional implications, leads to all those relativities in which being and validity are involved.

On this view of the evidence provided by intuition imperfection and even deception and illusion are possible, so that we cannot suppose we have finally arrived at the *truth* about some matter. This accords with our remarks on foundationalism in Chapter 1. The only way to correct what we obtain by experience is through further experience, through sequences of acts that make

our knowledge more determinate or in some cases even lead to radical readjustment. Thus in our experience we do not have truth as an absolute concept, but rather "truth" within the horizons of possible experience of an object as these are determined by the nature of the intention to the object at a given time. As our knowledge improves background beliefs associated with given intentions will become more determinate, so that we gradually converge toward truth. Now of course some areas of our knowledge are more determinate and secure than others so that the possibility of deception or illusion in these areas is minimal, perhaps even nonexistent. The evidence we can have for the existence of certain objects or statements, either general or singular, about those objects comes in different types and degrees. Husserl is concerned with a number of measures on evidence. The intuitive processes that produce a person's belief that a statement is true might yield apodictic evidence, adequate evidence, or perhaps even a priori evidence. Husserl also discusses degrees of clarity and distinctness of evidence, as well as degrees of adequacy. Another concern which should be noted is that for the intersubjectivity of evidence. It will be worthwhile to briefly consider these types of evidence now, leaving more specific considerations for later chapters.

There are a number of difficulties about trying to precisely characterize apodictic and adequate evidence, and the relation between them in Husserl's philosophy.[19] I shall not enter into these difficulties since it will not be necessary for my purposes. Instead I would like to simply take note of Husserl's mature characterization in *CM*. The kind of imperfection of evidence we have been alluding to thus far concerns the adequacy or inadequacy of evidence. Husserl says that with reference to it

> *Imperfection*, as a rule, signifies *incompleteness*, a one-sidedness and at the same time a relative obscurity and indistinctness that qualify the givenness of the affairs themselves or the affair-complexes themselves: i.e., an infectedness of the 'experience' with *unfulfilled components*, with *expectant* and *attendant meanings*. Perfecting then takes place as a synthetic course of further harmonious experiences in which these attendant meanings become fulfilled in actual experience. The corresponding idea of perfection would be that of '*adequate evidence*' – and the question whether adequate evidence does not necessarily lie at infinity may be left open.[20]

Husserl believes that much of the evidence we have is inadequate, and it is not clear whether he thinks there are any domains of knowledge where our evidence actually approaches adequacy. It will become clear in the next

chapter in any case that in ordinary perceptual intuition our evidence is not adequate. There are nonetheless degrees of inadequacy. We can usually improve upon the adequacy of evidence by carrying out further acts. In *EJ* Husserl notes that the stage at which one is prepared to break off further acts and to be satisfied with the evidence obtained will often be a function of one's practical concerns.

Apodicticity is another measure of evidence and according to Husserl's later views it can occur even in evidence that is inadequate. He characterizes it as a kind of "absolute indubitability". Much of our evidence will be inadequate, so that there will be a possibility of deception about the object and even the possibility that the object does not exist. This possibility can, he says, always be recognized in advance by critical reflection on the kind of evidence that we have. Apodictic evidence, however,

> is not merely certainty of the affairs or affair-complexes (states-of-affairs) evident in it; rather it discloses itself, to a critical reflection, as having the signal peculiarity of being *at the same time the absolute unimaginableness* (inconceivability) of their *non-being*, and thus excluding in advance every doubt as "objectless", empty. Furthermore the evidence of that critical reflection likewise has the dignity of being apodictic, as does therefore the evidence of the unimaginableness of what is presented with (apodictically) evident certainty. And the same is true of every critical reflection at a higher level.[21]

There are serious difficulties in coming to a clear understanding of this passage and of other remarks made about apodictic evidence. In just what kinds of cases do we have apodictic evidence, and thus apodictic knowledge? Husserl wishes to say for example that we have apodictic evidence for the existence of the self, or the transcendental ego, at any given point in time. Our evidence for the existence of the transcendental ego can thus count as apodictic. Even if we allow that this is so it is difficult to see whether we have apodictic knowledge about anything else, including the existence of other egos. Husserl's arguments on the latter question are notoriously difficult to follow. In *CM*, for example, he says at one point that other egos are not given with "unqualified" apodictic evidence, but he does not discuss the nature of "qualified" apodictic evidence. I do not intend to dwell on these questions but I will at least mention later some factors involved in determining whether our knowledge about natural numbers and finite sets could count as apodictic on Husserl's standards.

It should be pointed out that Husserl formulates the notion of apodictic

evidence for a state-of-affairs in terms of its being "absolutely unimaginable" that the state of affairs should not exist. Even if this condition is met it seems that we cannot suppose that we have some final, unimpeachable guarantee of the existence of the state-of-affairs. Apodicticity must also be viewed as a kind of regulative ideal that we can only approach. We might be incorrect in making some ascriptions of apodicticity. For practical purposes the notion of being "absolutely unimaginable" must always mean absolutely unimaginable to given subjects at given times and places, with particular powers of imagination, background beliefs, and so on.

A further question one might ask about evidence is whether it is a priori or a posteriori. Husserl suggests that there are a priori elements in both perceptual and mathematical intuition so that some of our knowledge must count as a priori. The question whether our knowledge in logic and mathematics is a priori has of course exercised philosophers for a long time. Husserl would no doubt claim that our knowledge in logic and mathematics is a priori but just what would be meant by this is not so clear. The usual view of the a priori is that it consists of what is contributed to our knowledge by the mind prior to experience. Husserl no doubt means to use the notion in this older sense but he also seems to refine the notion somewhat. Knowledge will count as a priori when what would be given in a sequence of intuitions is completely determined prior to actually carrying out the intuitions.[22] In this case we would have knowledge in advance of our actual experience. We would know how the experience would proceed and just what would be given at each stage in it so that we could "anticipate the experience". I shall argue later that reflection has a role in this knowledge. A priori knowledge on this Husserlian conception will be a function of background beliefs and the horizon associated with a given intention. In the case of certain types of intentions, especially in mathematics and logic, we evidently have a priori knowledge in this sense. I shall make a few more comments about this in later chapters as we take up some of the concepts relevant to understanding it better.

It is notable, finally, that Husserl is concerned with the intersubjectivity of evidence. This topic is discussed in some detail with respect to logic and mathematics in *FTL*.[23] For in Husserl's discussion of the "transcendental" foundations of logic and mathematics he is obliged to say that even in the knowledge of such sciences each mathematician M must start from his own "transcendental subjectivity". Having been obliged to say this the question of mathematical solipsism arises much as it does in Brouwer's writings on mathematics. Husserl at least says that solipsism is not really a problem in his philosophy but on the other hand it is well known that he wrote reams of

material on the problem of intersubjectivity. I would argue that there are not deep problems about intersubjectivity in Husserl's philosophy because fundamentally it invokes the Kantian strategy. The problems that do arise concern details about how to best describe intersubjectivity in the context of a theory of intentionality. For our purposes here it can simply be pointed out that evidence that is intersubjectively shareable is taken to be a standard to which scientific knowledge should aspire. Intersubjective agreement is surely a mark of the "objectivity" of knowledge. The lack of convergence toward intersubjective agreement signals an imperfection in our scientific knowledge. We might even argue that there could be no such thing as a "proof" or a "construction process" in mathematics that could in principle be understood by only one person.

We might now ask whether the possession of any of these types of evidence singly, or the possession of some combination of them, is sufficient to guarantee that we have arrived at the truth about the matters in question. For then we would be able to speak in an absolute way about *intuiting the truth* with respect to the area of knowledge in question. An answer to this question might be interesting for a number of reasons. An analysis of it could be expected to show under just what conditions it would be impossible for the sequence of mental acts through which objects appear to us to fail to provide the truth about some matter. As Tragesser has suggested, this would certainly have consequences for understanding the relationship between idealism and realism in Husserl's philosophy, and in fact for philosophy in general.[24] In mathematics it would give us a greater philosophical understanding of the relationship between our constructions and mathematical truth as platonistically construed.

On the basis of the view we have considered it seems that we could never in fact obtain the conditions that would be required. In the sequel I will especially concentrate on how adequacy would normally be out of the question. I shall speak of degrees of reliability and security of knowledge and be concerned to point out how in this regard there is an analogy between the evidence obtained by perceptual intuition and that obtained by number-theoretic and finite set-theoretic intuition.

7. COMPARISON WITH KANT

Husserl believes that the elements of a phenomenological elucidation of knowledge that we have considered represent an important advance on Kant's ideas on knowledge. As was mentioned earlier, Husserl's view of

knowledge as a product of (empty) intentions and intuitions bears an obvious resemblance to the Kantian idea of knowledge as a product of concepts and intuitions. Husserl however wishes to extend his distinction to logical and mathematical knowledge in a way that Kant did not. The acts involved in logical and mathematical knowledge have an intentional structure, just like any other acts. As such they are directed to objects and the intentions to these objects may be fulfilled, partially fulfilled, frustrated, and so on. Kant does distinguish between concepts and intuitions in mathematics, but Husserl points out that while "in Kant's thought, categorial (logical) functions play a great role ... he fails to achieve our fundamental extension of the concepts of perception and intuition over the categorial realm ... ".[25] Once we take account of acts directed toward mathematical objects it is necessary to investigate the constitution of such objects in consciousness. This is a kind of transcendental inquiry that Kant never undertook. Kant had no theory of abstract objects. Husserl suggests that perhaps Kant failed to ask transcendental questions about logic and mathematics because he was not in a position historically to fully appreciate the significance of the fact that logic and mathematics were about abstract objects. He was concerned only with the objects of sense-perception and with the constitution of the natural world. His failure to adequately account for logic and mathematics was largely a result of reacting to the empiricism of his time. The great progress in mathematics that resulted from the general move away from empiricist attitudes toward mathematics after Kant's time led to a fuller appreciation of the fact that mathematics was concerned with abstract objects or structures, and to the problems engendered by this view. This progress was attended by a much greater degree of formalization and abstraction in mathematics, especially in the late 19th century, and later by the clear refutation of psychologism. Husserl says that before this understanding was attained

> ... no one ventured ... to take the *ideality of the formations with which logic is concerned* as the characteristic of a separate, self-contained, *'world' of ideal Objects* and, in so doing, to come face to face with the painful question of how subjectivity can in itself bring forth, purely from sources appertaining to its own spontaneity, formations that can be rightly accounted as ideal *Objects* in an ideal 'world'.[26]

To the student of Kant Husserl's conception of mathematical intuition might sound like a form of "intellectual intuition" of the type that Kant dismisses. There is no reason to believe that this is true however since Husserl does not think of mathematical intuition as an activity or faculty that is independent of

sense experience. As noted earlier, Husserl explicitly says in many places that mathematical intuition must be founded on sensible intuition. The idea of a categorial intuition "cut off" from a "faculty of sensibility" is a piece of nonsense. Husserl believes that he is extending the kinds of views held by Kant.

8. INTUITION AND THE THEORY OF MEANING

Husserl's views on knowledge are connected with his theory of meaning in a way of which we should at least take note. It was said earlier that statements can express certain intentions and hence be meaningful even though we may not regard them as expressions of knowledge. Let us consider first how statements come to have meaning on Husserl's view. Printed words, utterances of sounds, and so forth are to be regarded as expressions of intentions.[27] A linguistic expression only has the *meaning* that it does because it is accompanied by the acts of consciousness that Husserl calls intentions. In the absence of such intentions language consists merely of meaningless marks of ink on paper, or utterances of sounds. Because meaning is primarily a function of intention Husserl tends to speak as if linguistic expression of intentions is not necessary for meaningful thinking. In the case of mathematics this suggests that linguistic expression is not essential for meaningful mathematical thinking, a position much like Brouwer's view that mathematics is "essentially a languageless activity of the mind". Of course it is necessary to be somewhat careful about this since much mathematical reasoning in particular is helped by the use of symbolic expression and would no doubt be impossible without it. In *Philosophie der Arithmetik* Husserl was especially concerned with this fact. But the particular form that linguistic expression takes is less important than the underlying intentions, for it is the intentions that one tries to communicate, however adequately or inadequately, in expression and that one tries to understand. In *LI* Husserl's portrayal of linguistic expression suggests that it has primarily a social function – it is necessary for communication and for scientific activity.

One might object to the view of mathematical statements as expressions of intentions on Fregean or Wittgensteinian grounds. For the view seems to imply that mental acts are the bearers of meaning and hence that, as Brouwer sometimes suggests, the meaning of mathematical statements is fundamentally private and subjective. But I do not see that the view needs to be understood in such a way that it is open to the kinds of objections raised by Frege and Wittgenstein. To say that mathematical statements are expressions

of intentions is not obviously inconsistent with the claim that mathematical language is public and objective, or that meaning is intersubjectively shareable. We would certainly not want to deny that a linguistic community exists in mathematics and the sciences. Since our primary concern here is with conditions for knowledge we might view this question of the meaning of mathematical statements in terms of a criterion of intersubjectivity of evidence for such statements. As we said earlier, evidence that is intersubjectively shareable should certainly be taken as a standard to which scientific knowledge should aspire, for intersubjective agreement is surely a mark of "objectivity" of knowledge.

Brouwer's view that mathematics was essentially a languageless activity of the mind, that mathematics was concerned with the precise part of our *thought*, was closely related to the dim view he took of formalization in mathematics. Brouwer was evidently opposed to formalization altogether. In this regard a phenomenological attitude would no doubt be different. It is well-known that Husserl was certainly not opposed to the effort to formalize theories provided that formalization was properly understood. His unpublished writings in particular show a strong interest in the role of formal systems in science. Nonetheless, formal systems have to be understood as simply more regimented, precise expressions of intentions. Viewed in this way, a phenomenological attitude toward formalization would be much like that of modern intuitionists: formalization is acceptable so long as it is not supposed that any particular formalization fully captures our mathematical thinking.

On Husserl's theory of meaning mathematical statements can express intentions and hence count as meaningful quite independently of whether they can be fulfilled or not.[28] In this respect Husserl's theory can be seen as something like the cognitive counterpart of Frege's view of sense and reference since on Frege's view expressions may very well have a sense but no reference. Husserl argues that statements may be objectless in principle, and yet they still have a meaning. One might show, for example, by an indirect argument that the assumption that there is an object satisfying certain predicates leads to a contradiction, but it does not follow that the statement has no *meaning*. It instead follows that it could not be fulfilled, or that there could not in principle be an intuition of the object. At one point in *LI* Husserl discusses the view that a statement should only be presumed to have meaning when it can be fulfilled.[29] On this view of meaning we only understand the meaning of a mathematical statement when we have or could have an intuition of the object or state-of-affairs to which it refers. Husserl ultimately

rejects this kind of account of meaning. In the process an interesting distinction is drawn between the meaning associated with an intention and the meaning associated with the fulfillment of the same intention. In rejecting the view that we only understand the meaning of a mathematical statement when we have or could have an intuition of the object or state-of-affairs to which it refers Husserl's view seems to differ significantly from Brouwer's since Brouwer actually says that certain mathematical statements, e.g., some statements of Cantorian set theory, are meaningless. In "Intuitionism and Formalism", for example, Brouwer gives a catalog of meaningless statements of Cantorian set theory. It is clear that the statements have no meaning for Brouwer because they cannot be obtained by what Brouwer calls the "basic intuition of mathematics" that he describes earlier in the paper. On a phenomenological view we should not be prepared to say that such statements are meaningless but only that at the present time there is no reason to believe that they can be fulfilled. Meaning is not to be identified with fulfillment, even though fulfillment of an intention may provide more information about the objects or states of affairs in question. In fact that is typically what constructive proofs do in mathematics. They provide quite specific numerical or computational meaning that is lacking in non-constructive proofs. The meaning associated with the fulfillment of an intention will be fuller and more determinate, perhaps more concrete.

Gödel's remark that mathematical questions not decidable now nonetheless have meaning and may be decided in the future seems to be in much the same spirit as the Husserlian view.

CHAPTER 3

PERCEPTION

1. INTRODUCTION

Most of the work that has been done by phenomenologists is concerned with the phenomenology of straightforward perception. It is in this area that one could expect to find the deepest and most illuminating analyses of Husserl's general distinction between intentions and the fulfillment of intentions. Perceptual intuition may be considered generally to be a process which provides the relevant evidence for a person's beliefs about medium-sized physical objects and about the physical world as a whole. Since the basic structure of the acts involved in ordinary perceptual intuition is supposed to be the same as that of the acts involved in mathematical intuition it would be a good idea to consider in more detail how the notion of fulfillment of intentions to ordinary perceptual objects is to be understood.

Husserl's theory of perception has been discussed extensively in the literature and I shall not attempt to do justice to its rich detail in only a few pages. I shall instead highlight certain features of a phenomenological theory of perception that are most relevant for the claim that perceptual intuition is analogous to mathematical intuition, and for the claim that perceptual acts are founding acts for acts involved in mathematical intuition. Since the interest in perceptual intuition here is motivated primarily by mathematical considerations I shall tend to gloss certain issues that might otherwise be treated in more detail. The most important features to be considered are structural features about the sequential nature of perceptual intuition: the "horizon" of perceptual acts; the temporality of perception; the feature of perceptual acts by virtue of which objects are individuated and identified; and the idea that intuition of an object as a whole is founded on the intuition of its parts (Recall this is only one of the senses of founding that we must consider.) I shall also comment briefly on the meaning of the widely misunderstood "phenomenological reduction" for a theory of perception since it would also be involved in a phenomenological account of mathematical intuition.

2. SEQUENCES OF PERCEPTUAL ACTS

A basic thesis of the phenomenological theory of perception is that objects are perceived in sequences of acts. Ordinary physical objects of course appear to us as individually identifiable, complete things even though what we actually see at a given stage in the perception of the object is only one of its parts or aspects. Perceptual intuition of objects takes place in successive stages in time and at any given time there will be parts of an object which are not actually perceived. In the perception even of something very simple, like a cup on a table, it is an essential part of the phenomenological description of this experience that not everything that could be seen or perceived about the cup is seen or perceived in one glance. The cup may have markings on the other side which I do not see, or its handle may be broken off, it may be unusable because it is cracked, it may have a rough texture on part of its surface and a smooth texture on another part, and so on. It is *possible* to make my knowledge of this object much more determinate by moving around it, or taking any of a variety of measures to become more thoroughly acquainted with it. The fact that this is necessary to improve my knowledge of the object is simply a consequence of the fact that at any given stage in my perception of the object it is not possible to see it in its entirety. Only parts of the object are intuited. The intuition of the object as a whole is founded on the intuition of its parts. I must have at least a finite sequence of perceptions of it if I am to gain any knowledge of it. It should be noted that at least under ordinary circumstances no one would deny that it is *possible* to make our knowledge of an object more determinate. Our reasons for believing it possible will be scrutinized later, for very similar beliefs are involved in mathematical knowledge.

Husserl describes knowledge of a perceptual object as a function of "chains of evidence" in which the same perceptual object appears. If the object does appear as individually identifiable, as an object in space-time, and as standing in some causal relation to us, then it must be possible to have a sequence of perceptions in which these facts are verified. One might suppose, for example, that at some stage in the process of perceiving the cup on the table the perceptual intuition breaks down. Suppose that what was perceived at some earlier stage of the process as a cup is seen at some later stage to be a paper cut-out that resembles a cup. Assuming that what was seen at earlier stages is not forgotten or distorted, and that later stages of perception verify the perception of the paper cut-out, then it is clear that pointing to the object at the earlier stage was not pointing at a cup. Just what it is that I am

ostending or individuating at a given time is to be viewed as a function of the sequences of perceptions of it. The only way to make adjustments or corrections in perception is through further sequences of perceptions. In extreme cases, such as in hallucinations, it is even possible to become aware of the fact that what was seen at some earlier stage to be an object in space and time, and one to which we stood in some causal relation, was really just part of an hallucinatory experience.

From the fact that an object is always perceived from a certain frame of reference it follows that at any given stage in the perceptual intuition of an object the object is underdetermined at that stage. Husserl discusses this in some detail in the *Ideas*, *CM*, and *LI*. In the *Ideas*, he notes that

> A certain *inadequacy* belongs ... to the perception of things, and that too is an essential necessity. In principle a thing can be given only "in one of its aspects," and ... that means incompletely, in some sense or other imperfectly ... *To remain forever incomplete after this fashion is an ineradicable essential of the correlation Thing and Thing-perception.*[1]

It is, in part, for this reason that objects of perceptual intuition are said to be "transcendent". The claim is not that the object itself is really incomplete or indeterminate, but rather that the *intuition* or knowledge of the object is incomplete. At any given stage in the perception of an object the knowledge of the object will be determined by the partial perceptions of it up to that stage. It is also the case that there will be associated with our intention to the object a certain "horizon of reference" in which the intuition of the object would become more determinate and complete. The horizon will consist of further possible sequences of acts in which the presence of the object or the presence of properties of the object would be verified. The horizon associated with an act will be determined by the nature of the act's intention at a given time and by the background beliefs that accompany the intention. Background beliefs will function to make the intuition more determinate. Previous experience will thus affect the present intuition. Moreover, it is not as if background beliefs are simply superadded to the data given at a particular time. Instead the background beliefs would be mutually modified and adapted to one another by the new incoming information. They would form what Gestalt theorists call a "functional whole". Certain structures and forms of organization come to be established in this way. Piaget, for example, has discussed this in some detail in his idea of how "schemes" of cognition come to be formed, established and consolidated. He suggests that from the point of view of empirical psychology such "schemes" be viewed as the functional

sensory-motor equivalent of concepts.

3. THE HORIZON OF PERCEPTUAL ACTS

Let us consider Husserl's idea of the horizon associated with an act in more detail. In *EJ* Husserl describes the horizon of an act in an interesting way. He says that

> Every act of experience, whatever it may be that is experienced in the proper sense as it comes into view, has *eo ipso*, necessarily, a knowledge and a potential knowledge (Mitwissen) having reference to precisely this thing ... this preknowledge is indeterminate as to content, or not completely determined, but it is never completely empty; and were it not already manifest, the experience would not at all be experience of this one, particular thing.[2]

Every experience can be extended in a continuous chain of individual perceptions which are united synthetically as a single experience. Husserl notes that depending upon my particular goals I may have enough of what an experience provides, so that I do not continue. But he claims that I can always convince myself that no individual perception of an object is the last, that what has been experienced still has, without limit, a horizon of possible experience of the same object. He says that " ... this horizon in its indeterminateness is copresent from the beginning as a realm (Spielraum) of possibilities, as the prescription of the path to a more precise determination, in which only experience itself decides in favor of the determinate possibility it realizes as opposed to others." Now what is especially interesting about Husserl's description in *EJ* is best appreciated by recalling that in *EJ* Husserl is trying to analyze the "origin" of logical and mathematical concepts in "prepredicative experience". For our purposes we may think of prepredicative experience simply as what we called straightforward experience or perception in the last chapter. Logical and mathematical concepts, as we said, are supposed to be founded on or have their origins in straightforward perception. In *EJ* Husserl describes the horizon of acts of straightforward perception as "the *induction* which belongs to essentially every experience and is inseparable from it, being in the experience itself." The term "induction" is used because "it suggests induction in the ordinary sense of a mode of inference and also because it implies that the later, for its elucidation to be completely intelligible, must refer back to the original, basic anticipation".[3] Husserl adds, still speaking of acts of straightforward perception, that

it is from the experience of anticipation associated with the horizon that "a true theory of induction must be established". Husserl is no doubt speaking here of a theory of empirical induction but I shall later suggest that there is an analog of this view for acts directed to objects like natural numbers and finite sets. Acts directed to these objects will also have their associated horizons. In the case of numerical or set-theoretic experience the horizon would determine just how we would extend our experience of these objects. Of course in this case we would have a kind of "mathematical" induction, not an empirical induction. A mathematical induction would nonetheless be viewed as an experience of anticipation with respect to constructions involving these kinds of objects.

A distinction is drawn between the *internal* and the *external* horizon of an act. The internal horizon consists of anticipations about the given object of an act, but the external horizon consists of anticipations or expectations about other objects of which we may be aware at the same time even though they are given in the background. The objects given in the background are objects toward which I am not actually directed at a given time but toward which I can be directed in further acts. As I take account of each object the object is in the foreground at that time, it is the object toward which I am actually directed. Husserl notes that the external horizon in perception also has an infinite, open-ended structure. In the example of the cup on the table the act was directed toward the cup. But of course I could also make the table the object of my perception, or any of a number of objects in my perceptual field. I could also make parts of given objects the objects of perception. The question of just how far I could iterate this latter procedure has of course been associated with some interesting questions in the philosophy of mathematics, e.g., about geometric intuition and continuity principles.[4] I shall not address questions about continuity in this essay but both the external and internal dimensions of the horizon will play an important role in my discussion of the intuitive content of a principle of mathematical induction in Chapter 6. The idea there will not be to consider perceptual intentions but rather intentions directed to natural numbers and finite sets. Acts directed to these objects have the same structure as acts directed to perceptual objects. In the case of mathematical objects like natural numbers and finite sets the nature of the "induction", that is, of the horizon associated with intentions to these objects, will be quite simple and well-determined.

In ordinary perception the partial perceptions of an object that are carried out up to a certain stage will play a role in determining what we could expect to know about the object at future stages in the sequence. These perceptions

provide a set of background beliefs about the object which we anticipate will be made more determinate as we continue to perceive the object. In normal perception earlier stages in the perception of an object which are nonetheless in the same neighborhood as the stage which makes up the present perception will actually be retained in memory and will not be distorted in any way. This kind of *retention* is discussed at length in Husserl's *PITC*. In cases where the sequence has a longer duration the actual retention of what was perceived at very early stages need not be part of the present perceptual act. However, it should be possible to recall what was perceived at earlier stages by an act of memory. In addition to retention there is also a protentive phase associated with each act. The retentive – primal impression – protentive structure associated with each act may be considered part of the immediate or local horizon of an act, as distinct from its general or global horizon which would involve a sequence of acts.[5] We shall discuss the retentive – primal impression – protentive structure in more detail in a mathematical context in Chapter 5.

There are several further points that should be noted about the horizon of an act. In the passage quoted a moment ago it is said that the horizon is "the prescription of a path to a more precise determination". This idea is also discussed in the *Ideas*. Husserl says that the horizon is "predelineated through the general meaning of the thing perceived as such, or by the general and essential nature of this type of perception that we call thing-perception." He continues by noting that the indeterminacy associated with the horizon of an act "necessarily means the *determinability of a strictly prescribed style*. It *points forward to* possible patterns of perception ... in which the continuously enduring thing in ever new series of perspectives reveals ever again new 'aspects' ... ".[6] Husserl says that we are referred of necessity "to unified and continuous series of possible perceptions which, starting from any particular perception, stretch out in an infinite number of directions in *systematic strictly ordered ways*, in each direction endlessly (without limit), and always dominated throughout by some unity of meaning". These passages confirm the view that the horizon prescribes in advance a process or procedure for carrying out the perception of an object or objects. Precisely how this is done is a function of the intention of an act at a given stage, of what is *meant* in the act. Not just anything could count as a possible further perception of a given object. In the example of the cup I expect that the cup could have any of a number of properties that are compatible with its being a cup. Precisely what its properties are will become more closely determined as I learn more about the cup. But it would be incompatible with my intending

an object as a cup, and with my belief that an object is a cup which is produced by my ongoing perception of the object, that it could be made of gears, flywheels, and pulleys, that it could be colorless, and so on. Certain possibilities are simply not compatible with my perception of the object as a cup.

The possibilities that are compatible are actually rule-governed or lawlike. Husserl states this explicitly in *CM* where he says that "*any object whatever* ... points to *a structure, within the transcendental ego, that is governed by a rule* ... the object indicates forthwith a universal rule governing *possible* other consciousnesses of it as identical ... ".[7] It is clear from other writings that the rules or laws which govern sequences of possible perceptions of objects can be quite general or quite specific depending on the category of objects under which the object is intended. Husserl of course believes that there are species/genus relationships between categories of objects that can at least be mapped out in particular intentional contexts. For the category of physical objects in general Husserl points out that

> ... the unseen determinations of things ... are, like thing determinations in general, necessarily spatial: this gives a lawlike rule for possible ways of completion of the unseen sides of the appearing thing; a rule that, fully developed, is called pure geometry. Further determinations of a (physical thing) are the temporal and the material; to them belong new rules for possible (thus not freely chosen) completions of sense and ... for possible ... appearances.[8]

For objects of different categories different laws or rules will be involved and these will be a function of just what is meant in intentions directed to objects of the given category.

It was mentioned earlier that the intuition of a perceptual object as a whole is *founded* on the intuition of its parts and it should already be possible to better understand this view. The sense of "founding" here and the underlying conception of parts and wholes are discussed extensively in the Third *LI*. To say that the intuition of an object as a whole is founded on intuition of its parts means that there could be no intuition of an object unless there were intuition of some of its parts. The intuition of the whole would in fact be inconceivable without intuition of its parts, for it is on the basis of the intuition of the parts that the various possible determinations of an object are fixed, marking it as a particular kind of object and not some other kind. Evidently we *know* more about an object than is literally given in some particular finite sequence of states. Were we to conceive of the states as

independent, discrete events this would perhaps not be true. But on a phenomenological view the states cannot be conceived this way. The states are in fact all interdependent. In ordinary perception a wealth of background belief will accompany each act fixing in advance a horizon of possibilities about an object. The way the object is perceived will be determined against the background beliefs. At a given stage in the finite sequence information will be present about future stages and information from past stages will also be retained. Our knowledge of the whole is founded on knowledge of parts in the sense of being determined, however inadequately, by what is given in the partial perceptions. The partial perceptions found our expectations or anticipations about an object or objects, they found the possibilities for further determinations. They therefore found our knowledge of the whole since some possibilities and not others are then fixed in a more or less determinate way. I shall not go into the matter but it is perhaps worth mentioning that writers like Hubert Dreyfus and Robin Gandy have noted that "Gestaltist" views of knowledge like this one may have important implications for the question whether there is a difference between human and machine intelligence.[9]

A further point about the horizon that is of some interest is that the sequences of acts associated with a particular act's horizon are to be thought of as sequences that we could carry out if we *actively directed* the course of perception in a particular way at a given stage. It is useful to note that where we are interested in obtaining more knowledge about an object or objects we would typically take an active attitude toward acquiring this knowledge. The particular way that we direct the course of perception will depend on just what we would like to know about an object, on our interest. Attempting to obtain knowledge about a perceptual object is often like attempting to solve a problem. We are interested in whether or not something is true of the object and the problem is to determine whether it is or not. In the simple example we have been considering, suppose I wish to maximize my evidence that the object I see on the table is a cup in some given amount of time. It can be assumed that I would check it in various ways: I would try to handle it, see it, and so on as completely as I could in the amount of time available. I would not simply stand in front of it for a certain period of time observing that it remained self-identical. I would have to take an active attitude toward the object. Thus the actual sequence of perceptions that I have will be determined to a certain extent by me and by what I wish to know. This does not mean of course that I determine what I shall see at each stage in my perception for that remains to be discovered in the experience, but it does mean that within the

range of possibilities determined by a particular intention I can typically make some choices about the course of the experience.

4. THE POSSIBILITIES OF PERCEPTION

In a number of the above passages it is suggested that the sequences of perceptions that we could have of physical objects are potentially infinite. They are moreover sequences that are ordered in time. The knowledge of an object or objects is acquired in successive stages in time. The idea of succession of acts in time is an essential feature of the phenomenological theory of perception. Of course the acts do not simply follow one another in time without any connections. Rather, what is given in later acts will depend on what was given in earlier acts. The results of previous intuitions will carry over into later intuitions. As a consequence of their successive character we should evidently think of acts of perception as linearly ordered, indeed as well-ordered. In considering the *possible* acts implied in a horizon however we would want to look at partially ordered structures of acts.

A consequence of the fact that the ordering of sequences of partial perceptions must be considered temporal is that certain interpretations of what is "possible in experience" must be ruled out. What is "perceptually possible", for example, will certainly have to count as different from what is logically possible on a classical conception of logic. But can we be more precise about the notion of what is perceptually *possible*? Husserl is certainly aware of the finiteness of human experience – we can in practice carry out only finite sequences of perceptions of objects.[10] Husserl claims throughout his writings, however, that for ordinary perceptual objects an "infinity" of partial perceptions would be required if we wished to maximize our knowledge of such objects. Objects of ordinary perception are capable of being perceived from an infinite number of different perspectives and the totality of these partial perceptions would, if they could actually be performed, characterize an object in all of its aspects. We can actually *complete* only a finite initial segment of any such sequence but by virtue of the act's horizon we would always know how to complete one more step in the sequence. We should, therefore, think of perceptual experience as potentially infinite. It is not possible for us to actually complete such a sequence even though we might be able to conceive of a being who could actually complete it. The intuitive powers of any such being would have to vastly exceed our own and evidently they would not be analogous to our intuitive abilities in certain crucial respects. For example, on the phenomenological conception of

the way that we progress toward maximal knowledge it would be absurd to suppose that a sequence that we could complete could be (epistemologically) equipollent to a proper subsequence of itself. But this would presumably not be absurd for the kind of omniscient being imagined by some philosophers and mathematicians.

Husserl's conception of the way that perceptual knowledge is acquired would evidently exclude some suggestions that have been made about the possibilities of perception. Suggestions about possibilities of perception that abstract from the incompleteability of sequences of perception involved in maximal knowledge, for example, or that abstract from the structure of temporality of perception would have to be excluded. Conditions about incompleteability and the structure of temporality, along with other conditions, presumably block accounts of intuitive possibility that have developed around Cantorian transfinite sets where this possibility is supposed to be understood as analogous to perceptual possibility.[11] Certainly for the very high infinities encountered in full set theory there would be very serious difficulties. It would be difficult, for example, to understand how it could count as possible to intuit structures of sufficiently high cardinality, say of $\mathfrak{I}_{\omega}$, whose existence is nonetheless uncontroversial for set theorists, since structures of such cardinality must far outstrip anything we could plausibly represent as obtained by a sequence of acts given in time.[12] It might be argued that these are just disanalogies between perception and mathematical intuition, but then it seems that perception would be at best a source of only quite remote analogies for mathematical intuition. In any case, I would say that on the basis of ideas of Husserl matters about transfinite sets are not clearly resolved for a number of reasons. One such reason is that statements about transfinite sets are not obviously objectless, as are some other kinds of statements. We cannot say in advance that we know that the intentions expressed by such statements could not possibly be fulfilled. Statements about transfinite sets lie in a gray area between statements that are provably objectless and, on the other hand, statements about objects for which we have elementary, sharply defined procedures for obtaining objects. Fortunately it is not necessary to pursue these matters here since I shall only be concerned with finite sets. For finite sets there will be, I think, a fairly direct analogy between possibilities involved in perceptual and mathematical intuition.

It might be objected that views about the possibilities of perceptual intuition that countenance sufficiently large finite sequences of perceptions as possible, or arbitrarily large finite sequences as possible, are already much too liberal. I shall return to this question once the reader has a clearer

understanding of just what epistemological consequences are thought to be bound up with it on a phenomenological view.

5. THE "DETERMINABLE X" IN PERCEPTION AND INDEXICALS

On the basis of his notion of the horizon of an act Husserl says that it is a task of phenomenology to

> ... make clear how the inadequate consciousness of giveness, the partial appearing, is related to one and the same determinable X, ... how ... a sequence of experiences is possible here ... , namely, the course of experience in which the empty places of the appearances that have preceded get filled again, the indeterminacies more closely determined, moving forward all the time toward *a thoroughgoing harmonious filling out* ... .[13]

In many passages toward the end of the *Ideas* Husserl speaks about sequences of acts in which the same object is given via the component of an act's intention known as the "determinable x". Each act in the sequence has its own x and in veridical perception of an object these x's are said to be "synthetically identified". If I attend to one object then the sequences of perceptions that I have of that object are all related to the same x. In the simple case of the perception of the cup each act in the sequence will be directed toward the cup. As I move around the cup, handle it, and so on, I perceive that this same cup has different properties. In the circumstance where I am perceiving different objects by considering them one by one, for example, each act in this process will be directed to its object by way of its determinable x. In this case the x of each act will correspond to something different and the x's in the sequence will not be synthetically identified.

Husserl's most detailed analysis of the function of the "x" component of an act is given in the *Ideas* in terms of the noetic/noematic structure of consciousness. Each act has a certain kind of structure consisting most simply, of thetic character, the Sinn component, and the "x" component of the Sinn. Various other features may be involved corresponding to modalities of giveness of objects, for example, but I shall not enter into these details at the moment. In the simplest case the Sinn component may be expressed in terms of a predicate F attached to an "x". Husserl says that in each act a property or properties is attached to a "something". The idea of the "x" is described in the following way:

> ... the predicates are predicates of '*something*', and this 'something' belongs together with the predicates ... it is their "bearer" ... We say that in the continuous or synthetic process of consciousness we are persistently aware of the intentional objects, but that in this experience the object is ever 'presenting itself differently'; it may be '*the same*', only given with other predicates, with another determining content ... There detaches itself as the *central noematic phase*: the 'Object', the 'objective unity', the 'self-same', the determinable subject of its possible predicates – *the pure x in abstraction from all predicates* ... .[14]

The function of this pure x has been rightfully likened to the function of indexical expressions in certain theories of reference. It is by virtue of the indexical character of the "x" that objects are individuated and identified. In the example of the cup, what a person M perceives at each stage is that this thing here is a cup. M has a perception *of* x, in this case, a cup. M will also typically perceive *that* this same thing, x, is red, has a broken handle, and so on. Perception *of* and *that* will both be involved in the perceptual situation.

Let us suppose, at least as a temporary measure, that in circumstances where we have sequences of perceptions of objects we express the x's of the acts at different stages 1, ..., n in the sequences as $x_1, \ldots, x_n$. We can then use '=' and negation to express that x's are or are not "synthetically identified" by putting, e.g., $x_1 = x_2 = x_3$, or $x_1 \neq x_2$, and so on. Then in the case where M perceives of some single object that it is an F, G, and H, we can express that the x's of the acts in which M perceives this are synthetically identified,

$$Fx_1 \wedge Gx_2 \wedge Hx_3 \wedge x_1 = x_2 = x_3.$$

Another possibility is that M perceives that

$$Fx_1 \wedge Gx_2 \wedge Hx_3 \wedge x_1 \neq x_2 \neq x_3.$$

Another is that

$$Fx_1 \wedge Fx_2 \wedge Fx_3 \wedge x_1 \neq x_2 \neq x_3.$$

Some details of this latter kind of circumstance will occupy us again in our discussion of induction in a mathematical context. It should be noted that other simple relations can also be expressed in this manner. M perceives that, for example, the cup is to the left of the saucer: M perceives that Lx_1x_2, and so on.

I have been purposely assimilating the "x" to the function of variables. The determinable x might be thought of formally as the component of the

intention expressed by a special kind of variable. This is rough and a much more refined analysis could no doubt be given. In fact proposals have been made by Smith and McIntyre which utilize possible world semantics and work by Kaplan and others on the semantics of demonstratives.[15] However, a working, intuitive understanding can be had which will make the analogy I wish to draw between perceptual and mathematical intuition readily appreciable. Let us think of the *content of the* x as just what it picks out or individuates on a given occasion. The x will have content only in some *context*, and apart from the context we will not think of it as having content. The use of the word "content" here fits Kaplan's use. In his discussion of "essentially occasional" expressions in *LI* Husserl distinguishes between the meaning that a demonstrative or indexical has in come context and the meaning that it has apart from any given context. The latter is called its *anzeigende Bedeutung* and the former its *angezeigte Bedeutung*.[16] In ordinary perception sequences of acts are carried out in a very definite context in which each x has its angezeigte Bedeutung. The context supplies the predicates that attach to the "x" making what it refers to concrete and determinate. We might thus also think of the anzeigende Bedeutung in terms of what Kaplan calls "character", although there are no doubt great differences between Kaplan's and Husserl's theories. To simplify matters and to bring out the structure of what is involved here we can think of the determinable x's as having content relative to some domain of individuals. Apart from any domain they still have an anzeigende meaning but no content. Hence their assimilation for formal purposes to the component of the intention expressed by something like individual variables. I wish to think of the structure of acts in such a way that this structure is fixed while underlying domains of objects may be varied. For later we shall observe the same structure where the objects involved are mathematical objects. Our notational use of "x" in the style of a variable should cause no special problems provided the underlying ideas are kept in mind.

In order to properly understand the function of the determinable x however it is useful to consider its role in non-veridical perception. In much ordinary perception our evidence is reliable. But it is of course possible to be deceived about objects. We already considered above one way that we could be deceived about an object. We might carry out some sequence of perceptions of what we believe to be a cup and then later we find out that *it* is a paper cutout that resembles a cup. Now it is the "it" here that the determinable x picks out. It is because the same object, the same "it" is involved here, as individuated by the x in each act in the sequence, that we realize that we have

made a mistake. The presence of something like an x in perception is a condition for the recognition of error. At various stages in the sequence we perceive that Cx and then at some later stage we perceive that $\neg$ Cx. Another way that we could go wrong in perception is with respect to the properties of an object that we do continue to perceive as the same object. We may believe, for example, that the cup is entirely red and then come to see that it is partly green. We can think of there being a set of predicates that are possible predicates of x, as determined by the act's intention. Only the ongoing intuition will determine which of these is to be believed true of x. In particular, it is only through the ongoing intuition that we have evidence for the intention that $(\exists x)$ Cx.

6. PERCEPTUAL EVIDENCE

Because an infinite number of partial perceptions of physical objects would be needed to maximize our knowledge the evidence that we have for the existence of such objects is inadequate. The kind of evidence that we have for the existence of perceptual objects cannot be considered indubitable. If we could maximize our knowledge of an object in a finite sequence of partial perceptions then clearly the kind of evidence in that case would be much more certain and secure. The best kind of evidence that we could have would be given in the case where no more than one partial perception would be required to maximize our knowledge: everything would be known about the object in one intentional act, and the object would not need to be seen from different perspectives. In this circumstance nothing would need to be made more determinate about the object and so there would be no question of having some further possible sequence of partial perceptions in order to completely survey the object. Because the evidence provided by perceptual intuition is inadequate we are entitled to speak only of degrees of evidence for the existence of objects. One basic measure on the degree of evidence is given by how far we have carried out the sequence of perceptions of a given object. One of the benefits of speaking of "degrees of evidence" for the existence of objects is that we need not feel committed to setting some definite bound on what should be considered "possible in experience", apart from some of the very general constraints mentioned above. For any bound we would set to mark some sequences as "possible" and other, larger sequences as not possible would certainly be arbitrary.

Since the evidence provided by perceptual intuition is inadequate different kinds of "counterevidence" are possible. One of the examples given earlier

about the perception of the cup showed how later perceptions in the sequence of perceptions of an object ran counter to, or "frustrated", rather than fulfilled the intention to the object as a cup. In *EJ* Husserl in fact describes the "origin of negation" in terms of the phenomenon of frustration or disappointment of intentions. Negation has its "origin" in the frustration of ordinary perceptual intentions in the sense that this phenomenon provides the immediate, non-reflective basis on which the awareness of negation is founded. The awareness of negation is originally constituted in the process in which an intention is frustrated.[17] There are different degrees of disappointment of an intention ranging from partial to complete. Husserl also discusses in this context questions about probability and improbability of fulfillment. In some cases negation will be experienced in the following way. Suppose that I have had some sequence of perceptions of an object and at each stage I saw that it was blue. I may form the belief that the object is entirely blue, based on my perceptions of it up to a certain stage. But suppose that after some further perceptions I see that it has a side that is red. This further perception "negates" my intending the cup as blue for now I see the cup is *not* entirely blue, but is rather blue and red. My continuing perceptual intuition has brought about a change or correction in my beliefs about the object. Even more radical negations and subsequent corrections may occur, as in the example where I even cease to see the cup as a cup. In the worst possible situations it can even turn out that my seeing an object like a cup was really just part of an hallucinatory experience.

In Husserl's writings a number of further observations about evidence become important once it is clear that the evidence provided by perceptual intuition will always be inadequate. Several measures apart from that of simply how far a sequence has been carried out are significant if our perceptual knowledge is to be as reliable and secure as it can be given the inadequacy. These measures will at least allow closer approximation to (absolute) truth and avoidance of mere "subjectivity" in perceptual beliefs. One of these conditions is that the process which produces a certain belief be one that I could repeat and in repeating it arrive at the same belief. Husserl says that the process that produces the belief should be one to which "I can return again and again". The procedure for determining that the object I perceive is a cup, if it is to count as evidence for the existence of the cup, should be one that I could carry out at will at different times. The idea that the process should be repeatable is of course to insure reidentification of the object. Being able to repeat a process in which an object is given permits reidentification and such reidentification must certainly count as a principal mark of "objectivity".

Hence, this condition is meant to distinguish cases of the actual presence of an object from cases of hallucinatory experience, and presumably also from cases of accidental attributions to the object. For in these latter cases there would evidently not be some repeatable process that would result in the same beliefs. In fact in the case of hallucinations there is usually not even a question of being able to return at will to some process that is supposed to present an object.

A process which gives evidence for a belief should be one moreover that everyone, or at least everyone with the appropriate skills, should be able to perform. As we mentioned in Chapter 2, intersubjective agreement will be important. The process may not be sufficiently clear to others, it may contain ambiguities, omissions, perhaps even a contradiction.

Knowledge of the lawlike or rule-governed character of the process will also have a bearing on the evidence we could be presumed to have. Apart form the fact that any such process will be a sequence carried out in time, we would presumably have better evidence for the existence of an object or objects when we could actually describe the sequence of perceptions in which an object is given by a rule or set of rules. We might have, for example, a set of instructions or a "program" which, when followed by anyone who understood it, would lead to a given perceptual belief. The rule or set of rules would give us a strict means for reidentification of an object or objects, since we would know a process that could be returned to at any time and which would, when carried out, always verify the presence of the object or objects. In this case our knowledge would be more determinate and less susceptible to doubt. If we could not describe the process by which we come to know something in this way then it might be possible to arrive at the same beliefs or knowledge again but we would not have a set of directions or a recipe for doing so. Surely our evidence in that case would not yield the same degree of security since certain decisions or choices in the process would be left to chance. In the case where we could describe processes by which we acquire knowledge by rules the condition for intersubjective agreement would presumably be satisfied since *anyone* could then follow the rule which results in a certain belief. This gives us more assurance of "objectivity".

7. PHENOMENOLOGICAL REDUCTION AND THE PROBLEM OF REALISM/IDEALISM

The views on perception that we have considered suggest, according to Husserl, that we think of perceptual intuition in terms of sequences of

indexical partial perceptions of objects and not in terms of some sort of special relation, causal or otherwise, and independent of this process, which is supposed to put us in touch with real objects. One of the classical philosophical problems with the latter view, of course, is how to explain cases of misperceptions, illusions and hallucinations, since there is, at least for hallucinations and some types of illusions, no real object with which one could stand in some relation. On the view suggested by Husserl the perceptual intuition of an object will be analyzed as a function of sequences of partial perceptions. This means that we never need to consider anything more than the sequences themselves, and the structure of the acts of which the sequences are constituted. The fulfillment of intentions to perceptual objects is to be explained in terms of such sequences or processes of verification: being able to have sequences of partial perceptions gives us not only the "concept of an object", but evidence for the existence of the object.

The phenomenological epoché or "bracketing of objects" could be understood in these terms. The bracketing of objects amounts to the fact that the consciousness of the "object" is just a function of the sequences of partial perceptions of it, so that we are not concerned with objects except as possible values of these functions. In phenomenology it is supposed to be possible to analyze the sequences of acts in which objects are given. Husserl says that "... Epoché with respect to all worldly being does not at all change the fact that the manifold *cogitationes* relating to what is worldly bear this relation *within themselves*, that, e.g., the perception of this table still is, as it was before, precisely a perception of this table".[18] The fact that various partial perceptions correspond to the same object, or to different objects, is a fact about these sequences, so the point is to discern and analyze principles that govern such sequences. I do not think that this position necessarily implies a commitment to philosophical idealism, or to some version of phenomenalism, but I shall not argue for that here. To do so would involve further discussion of the idea of the phenomenological/transcendental reduction and of the theory of intentionality. Husserl himself was evidently convinced that his position was compatible with some version of metaphysical realism, that is, with the view that the objects given by ordinary perceptual intuition are mind-independent.[19] The version of metaphysical realism would be different however from traditional, less sophisticated forms of metaphysical realism. Husserl argues that the physical objects we perceive are given as mind-independent in the very sequences of acts through which we come to know about them. Some of the premisses of the argument have already been indicated: the fact that the objects "transcend" our experience, the fact that

we cannot attribute properties to physical objects at will but rather must discover their properties – their properties appear to be fixed independently of us; the fact that properties of physical objects are given as different from properties of mental objects.

It seems fair to say however that an analysis of the sequential nature of perceptual intuition does not require that we take a stand one way or the other on the question of whether objects exist independently of the way that they appear to consciousness. In phenomenology it is not claimed that the existence of an object depends on the possibility of having a sequence of partial perceptions of it but rather that the intuition or *consciousness of an object as existing* does indeed depend on having such a sequence. To make this kind of distinction is a central concern in Husserl's attacks on psychologism.

A result of analyzing perceptual intuition in this way, as already mentioned, is that we do not need to consider the role of causality in perception. Husserl describes various structural features of consciousness which are necessary in order for there to be intuition of objects. Any more detailed analysis of these structural features, in particular of the sequential structure of perceptual intuition, would not require consideration of causal relations between physical objects and the sensory mechanisms of the perceiver. It could be argued that a phenomenological account would need to be supplemented by a causal theory in order to give an exhaustive analysis of the nature of perception. But if the idea is to analyze the sequential character of perception it would presumably be irrelevant to take up questions concerning causality at all.

CHAPTER 4

MATHEMATICAL INTUITION

1. INTRODUCTION

We are now in a position to launch more fully into a phenomenological account of mathematical intuition. We shall view mathematical intuition as a process which produces evidence for M's mathematical beliefs, much as straightforward perception is a process which produces evidence for M's beliefs about the physical world. We have already noted that from a phenomenological point of view mathematical objects are recognized to be of a different type from physical objects. Reductive enterprises that would seek to explain away mathematical objects are not viewed as true to mathematical experience. Acts of mathematical intuition and their objects are, however, said to be founded on acts of straightforward intuition and their objects. The intuition of mathematical objects will be founded on intuition of their parts, but mathematical intuition is also founded on straightforward intuition in the sense that the structure of perceptual acts provides the (representing) content for acts of mathematical intuition. Reflection and abstraction will be involved in acts of mathematical intuition.

On the phenomenological view of intuition it should be possible to set aside from the outset certain objections that have been raised to the notion of mathematical intuition. In the first part of this chapter I shall discuss objections about the analogousness of perceptual and mathematical intuition and about the view that intuition is "founded" on straightforward intuition that have stood in the way of attempts to account for mathematical intuition but that we could expect to avoid. Objections about analogousness can in part be deflected by observations already made: we claim that the analogy holds for the structure of the *acts* and the processes involved in intuition, but certainly not for *objects* of intuition. In mathematics we carry out processes where these processes are evidently about "abstract" objects. In perceptual intuition we also carry out certain processes, but these processes are about "concrete" objects. The processes involved in perception evidently contain probabilistic elements. The objects of mathematical and perceptual intuition are in fact disanalogous. Recognition of this fact will enable us to avoid some typical objections about the notion of founding as well. For we need not feel

compelled to represent mathematical knowledge as acquired from sources appropriate to properties of sensory objects but inappropriate to properties of mathematical objects.

After discussing these objections it will be possible to make a number of observations necessary for framing the account of intuition of natural numbers and finite sets that follows in later chapters. In particular, the basic logical apparatus underlying number theory and set theory will be introduced and interpreted in such a way that, in accordance with our remarks on empty intentions and intuitions, it does not abstract from intuitability conditions on objects. The general notion of a "construction" is introduced in this context and some further remarks are made about how we shall understand the notion of *existence* for natural numbers and finite sets.

2. OBJECTIONS ABOUT ANALOGIES BETWEEN PERCEPTUAL AND MATHEMATICAL INTUITION

There are a number of grounds for the frequently voiced objection that mathematical intuition could not have a role in mathematical knowledge analogous to the role of perceptual intuition in ordinary perceptual knowledge.[1] The grounds for the objection have a common source: mathematical *objects* like numbers, sets and functions are evidently abstract and therefore would have to be given to intuition in an entirely different way from ordinary perceptual objects. Ordinary perceptual objects admit of ostension, they are given as objects in space and time, as objects to which we are causally related, as having various sensory qualities, and as being immediately present in our experience. None of this could be true of mathematical objects. It is thought to follow, therefore, that there could not be a faculty of mathematical intuition analogous to that of perceptual intuition since mathematical intuition could have none of the basic marks of perceptual intuition. On a phenomenological account, of course, the conclusion is denied, but the premisses are perfectly reasonable. In fact an effort is made to show that mathematical objects could not have the characteristics of ordinary perceptual objects. Nonetheless, the structure of the acts and *processes* by which we obtain knowledge of mathematical objects is analogous in some important ways to that by which we obtain knowledge of perceptual objects. As mentioned in Chapter 2, intentions can be thought of as referring not only to objects that might exist independently of us but also to certain kinds of expectations about the course of our experience with respect to such objects. It is in the latter regard that mathematical and perceptual intuition will be

analogous.

It might be noted that the account described by Parsons in "Mathematical Intuition" is somewhat similar in recognizing the difference between perceptual and abstract objects. For sign types are abstract objects, albeit minimally abstract, but it is nonetheless argued that there is intuition of such objects. Concrete instances of the types, sign tokens, admit of ostension, are given as objects in space and time, as causally efficacious, as having sensory qualities, and as immediately present in our experience. The type is completely determined by what is or would be its tokens. It is the *instances* of the type however that are literally given in perceptual intuition. Although the relation of instance to object is looser for objects like natural numbers and finite sets we could hope for no more than that concrete "instances" or perhaps representations, not the objects themselves, would be available in straightforward perceptual intuition.

Of course we shall not want to say that intuition of a mathematical object is founded on sense perception of a concrete instance of it in the same way as in the sign token/type situation. For in the case of natural numbers and finite sets sense-perceptible "instances" will not serve as they do in the case of sign tokens and types to determine what objects the natural numbers and finite sets are. Numbers simply do not have concrete instances in the same way that sign types do, and finite sets are even farther removed from being intrinsically determined by concrete objects like sign tokens, especially when we begin to consider, for example, sets composed of sets. Of course we can think of numbers as generalized types as founded on "tokens" in a way analogous to the sign type/token case. But it is difficult to argue that on this conception we really get "the numbers". What is founded on concrete "tokens" in this case is really a particular construal of numbers that attaches to them properties that are alien to them as numbers. What this attests to more generally is that we do not want to think of natural numbers and finite sets as constituted by language as in the "numbers as generalized types" view, or the substitutional semantics view of natural numbers and finite sets. These views tie intuition to language or expression in a way that we shall not. On our view the function of founding and of constitution of the consciousness of numbers and finite sets shall be tied directly to the structure of cognitive acts and processes which, on a Husserlian view, is quite definite. The claim that mathematical intuition must be founded on straightforward intuition therefore means founding on the structure of previously given acts and processes of sense perception. We shall not think of the awareness of numbers and finite sets as constituted by perceptible linguistic expressions for such objects. We

shall of course want to develop this view in such a way that it does not imply psychologism or subjectivism about statements of mathematics, and to this end we follow the Kantian strategy.

The view that perceptual and mathematical objects are of different types, coupled with the claim that the structure of the acts and processes involved in intuition are analogous, allows us to avoid a variety of quandaries about mathematical intuition. We need not feel compunction to explain how causal relations with mathematical objects are possible since mathematical objects are not the kind of objects that stand in causal relations. The foundations of our knowledge of them are to be analyzed rather in terms of the mental acts and processes that produce evidence for our beliefs about them. Moreover, as has already been mentioned, the investigation of the structure of sequences of acts in which objects are given can be carried out quite independently of causal considerations, even in the case of ordinary perceptual objects. The claim here is not that ordinary perceptual objects are not given as standing in causal relations but rather that on a phenomenological analysis of our knowledge of ordinary perceptual objects our principal concern is with the structure of the sequences of acts in which objects are given.

We also do not need to be concerned about the fact that mathematical objects cannot be ostended or are not given as being in space and time, for mathematical objects cannot be expected to have such properties. The question of the immediate presence of such objects of intuition has already been mentioned in Chapter 2. Categorial objects are not given immediately in intuition but are rather said to be given "mediately" by way of being founded on underlying acts. The constitution of categorial objects takes place through active synthesis of the acts involved, not through passive synthesis in which an object is immediately present.[2] Ordinary perceptual intuition is "passive" in the sense mentioned in Chapter 2, so that the perceiver does not need to do or think anything in order to see an object. The object is presented immediately and to perceive it does not require undertaking any acts of collection, reflection or abstraction. The object does not need to be constituted through synthetic activities that must be carried out if there is to be awareness of the object at all. The awareness of mathematical objects however must be "constructed" in intuition.

To the possible objection that there could not be anything like mathematical intuition, analogous to perceptual intuition, because of the "passivity" of perceptual intuition, we could reply that there is a disanalogy in the sense that we clearly do not "construct" the sensory input in perceptual intuition. However, where we are interested in obtaining knowledge about an object we

can and usually do take an *active* role in directing our perception in order to determine what we wish to know. We can direct the course of our perception in different ways, by taking different perspectives on an object, handling it, and so on. The sequence of perceptions I have of an object, at least when I am interested in obtaining knowledge of it, is determined to a certain extent by me. I cannot maximize my knowledge of it if I do not undertake some procedure to become more familiar with it. In this sense we might even speak of "constructing" a perception of the object.

We also do not suppose on a phenomenological view that mathematical objects have sensory qualities like color, texture, and so on. In ordinary perception the sensory component of our experience falls under what Husserl calls the representing content of experience, and later in the *Ideas*, the hyletic data of experience. Hyletic data in sense perception are supposed to represent the various determinations that are made on our awareness of an object by sensory input, and this is an important component in the individuation of such objects. Hyletic data constrain perceptual intuition. We do not have the ability to change or rearrange aspects of a perceptual object or objects at will. Mathematical objects however do not have sensory qualities. The lack of sensory data in the intuition of mathematical objects does give rise to genuine problems for a phenomenological account of mathematical intuition. Parsons holds the view, for example, that Husserl lapses into obscurity in his attempt in *LI* to explain how in categorial intuition there is something analogous to sensations in sense-perception.[3] I agree on this point and so it becomes a question whether some sense can be made of Husserl's view. Another author has made a similar point in terms of what might be called the "plasticity" objection to mathematical intuition.[4] We can put the objection the following way. Perceptual intuition is non-plastic in the sense just described, that is, it is constrained by sensory data so that we cannot alter our perception of the object at will. This is obviously important for individuation of an object or objects since the sensory input will fix an object as a particular object and not some other. If mathematical intuition is not non-plastic in some analogous sense it is difficult to see how we could understand it as giving particular objects at all. In fact this problem is closely connected with issues about structuralism to be discussed in a moment.

A larger issue that lies in the background of the plasticity objection is that some such non-plastic component is needed in mathematical intuition to enable us to distinguish mathematical *intuition* from mere conceptualization or postulation.[5] It was noted in Chapter 2 that Husserl was already aware of this fact in *LI* where he points out that it is the representing content associated

with acts that constitutes the difference between empty and fulfilled intentions. Empty intentions may be directed to objects but they will have no representing content. Husserl therefore claims that in mathematical intuition there must be a feature analogous to hyletic data or representing content in sense perception, but of course it is some kind of nonsensuous analog of representing content. My approach to this problem will be to develop the analogy by focusing on the function that hyletic data have in perceptual intuition: they constitute a given which constrains or determines the acts of perceptual intuition. This is not to say of course that they fully determine what the intuition is of, for the same hyletic data can usually sustain different interpretations. We should therefore be able to distinguish a feature of mathematical intuition which constitutes a "given" that constrains and determines these acts in a similar way. I shall attempt to say more about this in Chapters 5, 6, and 7.

3. OBJECTIONS BASED ON STRUCTURALISM

The issue of individual identification and determinateness of objects of mathematical intuition should be discussed in more detail before we take up other objections. It has been argued that perceptual objects are individually identifiable and determinate but that mathematical objects are not. We could again point out that we recognize a disanalogy between perceptual and mathematical objects so that we cannot nor should we expect criteria of identity and individuation for one category of objects, ordinary perceptual objects, to carry over to a different category of objects, mathematical objects. This should certainly be true for the broad distinction in categories between ordinary perceptual objects and mathematical objects, and we should take care to heed it as well for distinctions within a given category, however these might be drawn. We could not, for example, expect identification of mathematical objects by ostension, nor could we expect to individuate them in terms of location, mass, color, and so on. On the other hand we would not expect to individuate ordinary perceptual objects in terms of arithmetical predicates like that of being prime, of being greater than four, and so on.

Having said that much we must still show that natural numbers and finite sets can be individually identified in acts of consciousness if we are to have an account of *de re* intuition of mathematical objects. It has of course been argued that we do not or could not intuit mathematical objects individually at all. Benacerraf has even argued that numbers (and by implication, finite sets) are not genuine objects at all.[6] These claims are accompanied by the view

that mathematics is really concerned with *structures* that model theories, never with individual mathematical objects taken singly. The properties and relations of mathematical objects that play a role in mathematical reasoning are those determined by the basic relations of a structure to which all the objects involved belong. Thus, the structures by which one characterizes mathematical objects like natural numbers, rational numbers, certain groups, spaces, and so on, determine all that there is to be determined about the objects in question. The objects characterized by the structure have no properties independently of those relating them to other objects of the same structure. In arithmetic, for example, any objects will serve as numbers so long as arithmetical operations are defined for them, and the laws of arithmetic are preserved. The properties of numbers which do not arise from the relations they bear to one another by virtue of being arranged in a sequence are of no consequence whatever for arithmetical purposes. One of the simplest examples that can be given about the indeterminacy of mathematical objects is the case of alternative set-theoretic construals of the natural numbers where it seems quite indeterminate whether the number 2, for example, should be identical to the class of all pairs, the one-element set $\{\{\emptyset\}\}$, the two-element set $\{\emptyset, \{\emptyset\}\}$, or none of these. It seems quite reasonable to conclude from such cases that we cannot think of the intentions expressed by number terms as being directed to some single object, or as having some single referent. Rather, there seems to be an indeterminacy with respect to just which object is picked out by any given number term. The problem that arises of course is whether this kind of indeterminacy is compatible with the claim that there is intuition of numbers. I shall argue that it is.

My argument will depend in part on some claims to be discussed in the next chapter about the undefinability of number. Husserl of course argued in his critique of Frege that the concept of number was not definable, and that one can only analyze the origins of the concept of number. His comments show that what he had in mind was that it was not possible to give an explicit definition of number, and surely not possible to give the kind of reductive definition in terms of concepts (or their extensions), sets, or classes, that leads to the problem of the indeterminacy of reference to numbers. It is nonetheless true that we easily make number determinations all the time, and that we have knowledge about number. So we might view the matter by saying that natural numbers and finite sets are objects that are "incomplete" or indeterminate with respect to reductive definitions. Let us concentrate on natural numbers. In the case of fulfillments of intentions to ordinary perceptual

objects some single object is fixed as the object of the intention in a particular perceptual context. Singular terms about perceptual objects expressing such intentions will thus appear to refer to some particular object. In the case of fulfillments of intentions to natural numbers some single object like $\{\{\emptyset\}\}$ will not be fixed as the object of the intention unless we have a particular mathematical context in mind as the background of our reasoning about numbers. In such a context a definite set-theoretic referent for the intention expressed by a number term would be supplied (that is, a referent as definite as the intention expressed by the number term in this context). The object would need to be supplied in a way that is unnecessary in perceptual intuition. But we cannot of course *identify* the number with this object. We can say that the object is playing the role of the number, or representing the number in this context. The relation of the number to the set-theoretic referent supplied in this context is that of object to representation. The number itself can be thought of, with respect to the set-theoretic definition, as "incomplete".

Should we be led to conclude from observations about the role of structures in mathematics that not only are numbers incomplete or indeterminate objects, with respect to the attempt to give reductive definitions, but in fact that they are not objects at all? Benacerraf, for example, argues that since numbers are not particular sets they are not sets at all, and then by extension of the argument, that since they cannot be identified with particular objects they are not objects at all. This kind of eliminative structuralism is too strong, especially since we do seem to have other grounds, to be discussed in the next chapter, for speaking of numbers as objects in at least *some* sense. Many additional difficult questions can be raised about the ontology of mathematical objects and I make no claim to have answers to them. We might wish to know, for example, whether mathematical objects are "objects" only in some sense derived from and secondary to that of perceptual objects. This kind of question about the "meaning of being" of mathematical objects is not something that I shall take up in any detail in this essay.

What seems to be the least objectionable position on the matter of reductive definitions of number can be taken by adopting the view, to put a twist on a phrase of Putnam's, that we think of numbers as *objects* of intuition if we like provided we keep in mind that each fact about them is a fact about any ω-sequence that would be generated by a given mental process.[7] We could therefore fall into line with the view of arithmetic as the science that elaborates the abstract structure that all progressions have in common merely by virtue of being progressions. It is concerned with all

structures of the order type of the numbers. But how is the indeterminacy of which we have spoken compatible with the claim that there is intuition of numbers? We would like to give, in effect, an account of intuition of numbers that does not commit us to any particular *construal* of numbers – that does not endow numbers with properties that are alien to them as numbers. To be a little more precise, we want an account that does not endow numbers with properties that are alien to them as *experienced objects* that are given as abstract. It might be objected that if we want an account that takes numbers as experienced objects we could just as well settle for the "numbers as generalized types" view. The problem with that view however is that it makes experience of the object – a number – depend on straightforward sense perception and perceptual representability is an extra property that we do not want to attribute to numbers. Our account should be compatible with structuralism plus the view that numbers are nonetheless objects in some sense, not with structuralism plus the view that numbers are not objects at all.

To continue our argument we shall note that we can speak of individuation of numbers in acts of consciousness in the sense that we can determine their place in a sequence in relation to one another.[8] We can agree completely with Benacerraf's remark that "One cannot tell what number a particular expression represents without being given the sequence of which it forms a part. It will then be from its place in that sequence – that is, from its relation to other members of the sequence and *from the rules governing the use of the sequence in counting* – that it will derive its individuality".[9] Our claim that there is intuition of numbers depends on just this kind of observation. For as we shall see in the next chapter, number intuition is a process which is to be thought of as giving the sequence of which a particular number forms a part. The function of intuition of numbers is to determine in a rule-governed way the position of numbers in sequences. It is in this sense that we have the means for individually identifying the objects to which the intentions expressed by number terms refer.

On our view of the intuition of natural numbers we will have a very natural explanation of how we could know that, for example, $7 \neq 5$ or $6 + 3 = 9$, or that $5 < 6$. For intuition gives the means for individuating the objects referred to by the singular terms here. It should be noted that our knowledge that singular statements such as these are true or false does not require at all that we know just which particular set-theoretic objects the numbers are, e.g., Zermelo sets, or von Neumann sets, or any of the other possibilities. In fact to ask for such identification is beside the point – it is really overdetermination for no particular reason in the face of the question whether such statements

are true or false. The account of intuition of numbers that we shall give explains in a reasonable way the knowledge that we have about the truth or falsity of the statements about numbers. We shall understand fulfillment of an intention expressed by a number term as a sequence of acts in which the number is given by virtue of its place in the sequence. This accords fully with our emphasis on thinking of intuition as a *process* in which objects are given, or on viewing intentions as referring not only to objects thought to exist independently of us but also to processes in which such objects would be experienced.

The details of the view of the intuition of numbers needed to support our argument are gone into more fully in the next chapter. It might be noted from what has been said, however, that we do seem to have grounds for speaking of numbers as objects in some sense. The mere fact that we speak of quantifying over natural numbers should already be taken as grounds for speaking of numbers as objects in *some* sense. In the next chapter we shall argue that we can individually identify numbers in sequences in which they are given and that this kind of individual identification should surely count as a criterion of objecthood. We might think of numbers as objects of intuition in the sense that they have a place in the sequence in which they are generated. This place can be reidentified and we can formulate identity conditions for it.

Remarks about the individuation of finite sets involve some observations not made here but a similar outlook can be had if we think of a theory of HF sets formulated in a way quite analogous to arithmetic. We shall think of finite sets as also generated by a certain process. Questions about finite sets are taken up in Chapter 7.

4. OBJECTIONS ABOUT FOUNDING

Let us now consider some specific objections that have been raised to the thesis that mathematical intuition is founded on perceptual intuition. This thesis can be construed in quite different ways. What is perhaps the most common construal amounts to supposing that what we know about mathematical objects like numbers is determined by perceptible representations of the objects. In the case of numbers and finite sets the most obvious candidates for being perceptible representations of these objects are *concrete sign configurations*. Hence this construal has accompanied the view of mathematical intuition as being of or determined by concrete sign configurations. This way of construing the thesis, and the notion of mathematical intuition that goes with it, is often attributed to Kant and to Hilbert.[10] In precisely what

form they actually held it and with what qualifications, is not clear to me. In any case the view is open to a number of serious objections that are worth considering if for no other reason than to more clearly demarcate our own conception of mathematical intuition. From what has been said it should be obvious that we construe the thesis in a different way.

One objection is that we could not have mathematical knowledge if this knowledge depends on intuition of finite sign configurations because sign configurations have properties that are irrelevant to the mathematical objects we are supposed to be considering. If we view the sign configurations in terms of sign tokens then they have properties like that of being written in a certain kind or color of ink, occurring at different times and places, being written by different people, and so on. Viewing the sign configurations in terms of sign types will not help because sign types still have properties quite different from the properties of mathematical objects like natural numbers and finite sets. In either case the properties of the sign configurations would be irrelevant. But if our arithmetical or set-theoretic knowledge depended on such representations how could we know that these properties are irrelevant? The knowledge that the properties are irrelevant could not depend on our perceptible representations. Whatever ability or process allows us to distinguish between relevant and irrelevant properties for natural numbers and finite sets would not be due to mathematical intuition (on this conception of mathematical intuition) but would have its source elsewhere. It is concluded therefore that mathematical intuition, as understood on this kind of account, could not be what provides arithmetic or set-theoretic knowledge. It does not give the kind of evidence appropriate to knowledge of mathematical objects.

Several other objections closely related to this one can be developed. It could be objected, for example, that we could not know that our sign configurations for natural numbers and finite sets were accurate if this knowledge depends on a view of mathematical intuition according to which the objects of intuition were finite sign configurations. Consider, for example, the sign configuration '||||||' which we might take as a perceptible representation of the number 6. How could we know that this is an accurate representation of the number 6? We could not refute the claim that it was not an accurate representation by asking a person to look again. For nothing we could learn about the *sign configuration* '||||||' with the properties *it* has would convince us that it accurately represented the number 6. The knowledge that it does accurately represent the number 6 must therefore depend on some other source of mathematical knowledge apart from intuition that is of or is determined by sign configurations. From the point of view of

phenomenology we could certainly agree with arguments of this kind. For we do not wish to suppose that what we know about natural numbers and finite sets is determined by sign configurations for these objects. We shall not suppose that mathematical intuition is founded on straightforward intuition in that sense. Certain other objections in a similar vein can also be set aside. We do not need to worry, for example, about whether there is an upper bound on the number of physical inscriptions in the universe since mathematical knowledge will not be founded on the availability of such objects. The proper conclusion to draw about the sense in which one could say that arithmetical and set-theoretic knowledge is "founded" on the intuition of concrete sign configurations would seem to be that suggested by Parsons: this conception of mathematical intuition gives objects that can be taken as elements of a model of arithmetic, and possibly of a model of finite set theory, and hence could be taken to guarantee the existence of elements of at least one model of these theories. It might then be argued that without this kind of intuition we could not have such a guarantee, and we would perhaps not therefore have a guarantee that a structure satisfying the principles of arithmetic exists. But there is no intuition of natural numbers here.

Objections about founding on concrete sign configurations apply not only in the case of relatively small natural numbers and finite sets and singular statements about them but perhaps even more forcefully in the case of large numbers and in statements proved by mathematical induction. Let us focus on statements proved by mathematical induction. It can be objected that insofar as mathematical intuition depends on our sign configurations for mathematical objects any attempt to argue that the knowledge we have by induction is founded on intuition will be circular: it will presuppose the very knowledge that intuition is supposed to make possible. For in this case the knowledge we have by induction would depend on seeing that induction holds in particular instances where we substitute numerals, i.e., perceptible representations of numbers, for variables. Let our numerals be strings of strokes. Then, for example, if we know that $S(|)$ and that $S(|) \rightarrow S(||)$, then by modus ponens we know that $S(||)$. We might continue in this style but it will not enable us to se that induction holds as a general principle. To see that it holds in instances does not amount to knowing that it holds for *every* numeral. To claim that it does show induction holds for every numeral would be circular: the only way we could be convinced of that would be by induction. It will not do moreover to argue that the knowledge we have by induction is founded on intuition of sign configurations because we can represent reasoning about "arbitrary" numbers in intuition. To represent

reasoning about an arbitrary number we would presumably need to consider an arbitrary string of strokes like '| ... |', where the '...' stands for some indeterminate number of strokes. If we take '| ... |' to have some particular number of strokes it would not be arbitrary and hence would not enable us to see that induction holds for *every* numeral. On the other hand if we do not suppose that '| ... |' contains some particular number of strokes we could not see its internal structure that we would literally need to see in order to know that induction holds for it. We can therefore conclude that the knowledge we have by induction is not founded on mathematical intuition if mathematical intuition depends on sign configurations as representations of natural numbers.

If mathematical intuition is not to be understood as primarily of or determined by concrete sign configurations, as is suggested by our view, we might still hope for an account of how the knowledge we have by induction is founded on intuition. The objection we have considered is less a problem for our account than it is for a Hilbertian or possibly a Kantian view of intuition. There are however other problems about intuition and induction that do not depend on the supposition that intuition is of or is determined by sign configurations. There is a problem, for example, about how we could know that conclusions by induction are true since to see that they are we would need to have a sequence of intuitions that we could not in practice carry out. There is evidently a finite upper bound on the number of intuitions that a person could carry out. The usual reply is that we at least know that "in principle" a sequence of intuitions of the required sort could be carried out. But how could we know that such a sequence could in principle be carried out if we are not able to actually carry it out? One might be inclined to argue that either we could not really be said to know it, or that if we could be said to know it this knowledge could not depend on intuition. We shall have to consider this problem in more detail in Chapter 6.

From earlier remarks it is perhaps obvious how we shall understand the claim that the knowledge we have by mathematical induction is founded on intuition. The cognitive processes involved in intuition have a quite definite structure. We shall want to see how this structure is involved in founding knowledge of singular statements about numbers and finite sets, and how it is also involved in founding knowledge of general statements that we have by induction. We began to mention in Chapter 3 processes involved in perceptual intuition upon which our knowledge about large numbers and finite sets, and our knowledge of conclusions by induction would be founded. In particular the structure involved in the horizon associated with intentions

directed to natural numbers and finite sets should be quite definite. The underlying experience in perceptual intuition upon which knowledge that we have by mathematical induction will be founded is the anticipation or expectation associated with an act's horizon. As we said, an act's horizon fixes possibilities about the future course of experience as a function of the intention of the act and various background beliefs. This will be true for intentions directed to objects like natural numbers and finite sets just as it is true for intentions directed to other kinds of objects. It seems to me that by emphasizing the view of intuition as a *process* with the various structural features noted we shall have a distinct advantage over accounts of the intuitive content of mathematical induction that leave out facts about intuition as a process, or as a sequence of acts. I have of course given only a very brief statement of how we shall approach questions about the intuitive component of proofs by induction. Details and difficulties will be considered at various points in Chapter 6.

5. A LOGIC COMPATIBLE WITH MATHEMATICAL INTUITION AND THE NOTION OF CONSTRUCTION

Let us now turn from this discussion of some of the main objections to the notion of mathematical intuition to make some observations about basic logical notions that will underlie our treatment of both number-theoretic and set-theoretic intuition. We shall frame an account of mathematical intuition in which these notions play an explicit part. We could think of a "logic-free" development of arithmetic and of arithmetic notions from which we could define logical operations, but it seems preferable to take logical operations as primitive since this is a widespread practice in foundational developments, since we could think of different mathematical theories in terms of different extensions from this foundation, and since, in particular, we can express statements about the existence of objects in a regimented way with existential quantifiers so that our talk of the existence of objects will be plain to those for whom it otherwise might not be.

We would like our understanding of the logical operations that would occur in mathematical statements to reflect the view of knowledge that has been discussed. Knowledge is to be viewed as a product of intentions and intuition. We shall view statements from different kinds of discourse as expressions of intentions. Thus, mathematical statements are to be viewed as expressions of intentions. It should be noted that there are good reasons for thinking of expressions of intentions as always being expressions of a certain

type, in accordance with something like a theory of types. The reason for this is that an intention can itself be made the object of a cognitive act, and this cognitive act would be directed to the intention (as object) by way of its own higher-type intention. Cognitive acts which have intentions as their objects are acts of reflection. In ordinary experience acts are directed toward ordinary kinds of objects, not toward intentions. But reflection is always thought to be possible, and this should be taken into account for mathematical purposes. Hence one gets very naturally the idea of a possible hierarchy of intentions that should be reflected in intensional mathematics. I shall not pursue this kind of development in any detail in this essay but it plays an explicit role in theories like those of Martin-Löf.

We shall want to interpret logical operations in such a way that we do not abstract from conditions of fulfillability of intentions, that is, from conditions of intuitability, since knowledge is to be viewed as a product of intentions and intuition. As we have been saying, intentions can be thought of as referring not only to objects that might exist independently of us or of our constructions, but also to *processes* in which the presence of objects would be verified. Associated with intentions directed to particular objects will be certain kinds of expectations or anticipations about the course of our experience with respect to the objects. The horizon of an act indicates a course of possible experience in which the intention to an object would come to be fulfilled, partially fulfilled, frustrated, and so on. From this point of view an "empty" intention is a pure expectation about an object, and it may be either "realized" or not. In our treatment of logical operations therefore we shall adopt the position suggested by our earlier comments, and first offered by Oskar Becker and Heyting, of explaining the meaning of logical constants in terms of the notion of fulfillments of intentions.[11] Heyting of course identified the notion of a mathematical *construction* or *proof* with the fulfillment of the intention expressed by a mathematical statement: We have a construction for a mathematical statement S iff the intention expressed by S is fulfilled (or is fulfillable), and we lack a construction otherwise. We can think of ourselves as adopting the term "construction" in the case where we are speaking of the fulfillment of intentions expressed by mathematical statements in particular. This is especially useful for conveying that the intuition of mathematical objects must take place mediately through active synthesis on the basis of underlying, pregiven acts or processes. Mathematical objects must be "constructed", or "realized" in a way different from objects of straightforward intuition.

If we think of empty intentions as pure expectations we might reformulate

the notion of a construction by using Kolmogorov's explanation, according to which a mathematical statement is to be viewed as the expression of a problem. Then the intention expressed by the statement is fulfilled iff the problem is solved. Sundholm gives an argument, attributed to Martin-Löf, to show that the Heyting and Kolmogorov explanations are substantially the same.[12] In a recent paper in which an analogy between constructive mathematics and computing science is drawn Martin-Löf makes a related observation.[13] In constructive mathematics we need to have an understanding of "a is a construction for S". If we identify the processes (functions) involved in the construction with programs, and use the Kolmogorov explanation, we can understand "a is a construction for S" as "a is a program for the problem (task) S". "Programs" are, so to speak, processes for fulfillment of intentions.

This last remark raises an important issue for the work in this study and we should say something about it now, postponing a more detailed treatment of it to another occassion. In his comments Martin-Löf is evidently thinking of *machine* programs, but we should perhaps think of them, as Bishop has put it, as "person programs". The question whether this amounts to a significant distinction raises some large issues about whether there is a difference between machine and human computability. Consider, for example, the following function "defined" on Gödel numbers of formal derivations of HA (due to Kreisel):[14]

$$f(n) = \begin{cases} 0 & \text{unless n is the Gödel number of a derivation in HA of a closed sentence of the form } (\exists x)\, P(x), \\ 0 & \text{if derivation n does not provide a \textit{specific} numerical instance k satisfying } P(x), \\ k+1 & \text{if k is the specific instance of } P(x) \text{ provided by n.} \end{cases}$$

Because we are speaking of HA from the point of view of intuitionistic philosophy the definition of this function involves reference in each case to the mental act of which the formal derivation with code n is the expression. Coding the mental states of the human computer may be a more delicate matter than coding the states of a Turing machine. In the case of Turing machines, whose states are finite spatio-temporal configurations, it is quite clear how to code the states by natural numbers. For machine computability the computations and their arguments are on par. But, as Kreisel has suggested, human computations may be more "complicated" or more "abstract" than the objects on which they operate. Kreisel contrasts this with a formalist philosophy which requires that our reasoning be represented by formal

derivations. The latter view in formal axiomatic set theory would, for example, make our "thoughts" *less* complicated than the objects, the infinite sets we think about.

The *phenomenon* which suggests that machine computability is different from human computability is simply that a Turing machine will not react to f as it stands, and yet we evidently know how to execute this function. We are required first to decide whether a derivation n provides an instance k, and secondly, to find the instance. There are various recursive rules for extracting an instance k such that P(k) from a derivation n, but there is no specification here about which particular method should be followed. The function evidently defines for *us*, as it stands, a sequence of integers. Moreover, its values would not be generated by a random process, or by arbitrary choices. The human who has to execute this process is evidently free to use his or her creativity to extract an instance from the proof. As it stands, it is apparently humanly computable (also constructive), perhaps a "creative subject" computation, but not machine computable. A rough view of what we might say phenomenologically about the function is that, on the analogy with perception, a mathematician M sees it as definite and knows how to work with it even though, strictly speaking, he is not given all its "parts" or "aspects". Not all of its "parts" are explicitly given, just as ordinary perceptual objects are literally given in an incomplete way. Thus M must bring to bear various background beliefs in order to execute f, and the exact nature of this process may involve a complexity not involved in a Turing machine computation. But the function would have to be given in a very explicit form to elicit behavior from a Turing machine. We shall not pursue this matter further here however.

It should be noted that in identifying constructions with fulfillments of intentions we do not wish to place constraints in advance on the type and degree of evidence required for the mathematical intention expressed by S to count as "fulfilled". Such constraints might determine different styles and grades of constructivism depending on what one wishes to require of the evidence involved. Different mathematicians would no doubt draw the line on constructivity at different places. Compare, for example, an ultrafinitist conception of what is constructive with that of a predicativist. I shall not try to make any definite statements about the limits of constructivity. Apart from some very general features that distinguish constructive from non-constructive reasoning, e.g., the general use of pure "existence" proofs and excluded middle, it seems we should recognize that in mathematical practice constructivity can be construed in different ways. We should not expect the universe

of constructions to be clearly defined in advance. We can say that at one extreme we have intentions expressed by some mathematical statements for which it is impossible to find constructions for obtaining objects (eg., largest prime numbers, roundsquares), and at the other extreme that we have intentions for which there are very elementary, clear-cut construction procedures. Examples of the latter type would include procedures defined by rudimentary predicates or functions, by Kalmar "elementary" functions, or by primitive recursive functions.

In considering questions of evidence one should no doubt take issues about the feasibility and complexity of computations seriously. But it might still be argued that between these extremes there are gradations of constructivity. It is in this area that one encounters difficulties. Should we say, for example, that because we have a "quasi-combinatorial" understanding of how transfinite sets are obtained in the cumulative hierarchy that these objects in some sense belong to the universe of constructions? That would no doubt be a *very* liberal interpretation of constructivity. But are statements about transfinite sets obviously objectless, are they not obtained by a kind of "procedure"? Evidently the procedure is one that only a god could carry out, not us. At least for elementary constructions we might regard these differences as differences in requirements on the degree of evidence we have for objects. There are grades of evidence for objects and so there are grades of knowledge. Intentions can be not only completely fulfilled, but also partially fulfilled, problems can be partially solved, and this must surely count as giving us *some* knowledge about the object or objects involved. Other factors concerning evidence may also play a role, e.g., presence or absence of lawlikeness, finite bounds, predicativity, intersubjectivity, etc. It should be noted, incidentally, that in the case where the intention expressed by a mathematical statement is completely frustrated we also have knowledge. As Heyting puts it, negation is something "thoroughly positive". Since I shall be speaking only of natural numbers and finite sets my comments in later chapters on constructions and evidence will be specific to these objects. I do not think that identifying constructions with fulfillments of intentions by itself forces any particular version of constructivism, but our position is perhaps closest to that of traditional intuitionism with its emphasis on mental acts, reflection on mental acts, the role of thought in relation to language, and so on.

It should be obvious that our position also differs in some respects from that of traditional intuitionism as represented by Brouwer. I do not wish to go into this here but I would like to briefly mention one matter. We said that we

shall explain the *meaning* of logical constants in terms of the notion of fulfillments of intentions. We are therefore understanding the notion of meaning here in the sense of "meaning as fulfillment". We do not wish to deny that mathematical statements have meaning because there is no reason at present to believe that they can be fulfilled (or frustrated). In "Intuitionism and Formalism" Brouwer says that many such statements are meaningless (for example, statements asserting the existence of nondenumerable totalities), and perhaps some modern theorists of meaning would follow him in this. There are indications elsewhere in Brouwer's writings, however, that he did not mean to adopt this view.

If we suppose that we have an understanding of the statement that "the intention expressed by a mathematical statement S is fulfilled" we can give a number of clauses indicating how constructions for composite statements are to be understood. We are of course taking the notion of construction here in a very general sense. It seems to me that in particular mathematical contexts we can have a very definite idea of what it means to say that the intention expressed by a mathematical statement S is fulfilled, but that our general conception of this is not so clear or definite. We do not have a clear or definite understanding of the totality of all constructions. Because of the nature of our knowledge it seems unreasonable to expect that we should have. We shall at least try to come to a better philosophical understanding of what it means to say that the intention expressed by a mathematical statement S is fulfilled in the case of statements of elementary number theory and finite set theory. With this in mind we shall understand logical constants as follows:

(i) A construction for $S \wedge T$ consists of a construction for S and a construction for T;

(ii) a construction for $S \vee T$ consists of a construction for S or a construction for T;

(iii) a construction for $S \rightarrow T$ consists of a construction which converts each construction for S into a construction for T;

(iv) a construction for $\neg S$ consists of a construction for $S \rightarrow \bot$, where _ is some false or absurd intention like $1 = 2$.

Let us assume for the quantifiers that some domain of objects D is given. We want to think of D as constructible.

(v) a construction for $(\exists x)\, Sx$ consists of a construction of an object $a \in D$ and a construction for Sa;

(vi) a construction for $(\forall x)$ Sx consists of a construction which associates to each $a \in D$ a construction for Sa.

The quantifier clauses are stated in a general way but we shall consider two specific domains of objects and some of the philosophical questions associated with supposing objects of the domain are given.

Formal or mathematical interpretations based on this kind of explanation have of course been developed extensively. As we are looking at the matter it is best to think of the clauses as providing a philosophical view about constructions and not what we would now think of as a *mathematical* theory of constructions analogous to Tarski's mathematical theory of truth. Examples of mathematical interpretations based on this explanation are those utilizing various forms of realizability and Gödel's *Dialectica* interpretation. The mathematical interpretations give a quite specific and more precise meaning to the notion of "a construction for S" which is amenable to mathematical treatment. Attempts to develop more directly a formal theory of constructions for constructive mathematics on the analogy of Tarski's theory of truth for classical mathematics have of course been undertaken by Kreisel and Goodman.[15] Scott and more recently Martin-Löf have also developed what may be considered theories of constructions. J. Smith has developed theories like those of Martin-Löf in which logical operations are taken as primitive.[17]

I shall not comment on these theories except to note that a point that Sundholm makes about the Kreisel/Goodman theories should be obvious from our remarks.[18] Namely, that the "proof predicate" Π (x, S) in the Kreisel/Goodman theory should not be viewed as a mathematical proposition (or statement) on the same level as S itself. The relation between a statement and a construction that fulfills the intention expressed by the statement is not itself statement-like or propositional. Rather the fulfillment of an intention is to be understood *as an experience* that is or can be achieved. Heyting was already careful to stress this point. If we were to view the relation as a statement we would find ourselves involved in some rather odd consequences for then we would have to view it as expressing an intention toward yet another process of construction, thus starting a vicious regress. We shall of course not suppose that we think of the relation as expressing an intention toward yet another construction process for the same reasons that we would dismiss such a view as absurd in the case of ordinary perceptual knowledge. I believe this kind of observation should help to provide a positive answer to the question raised by Dummett and others of whether it is possible for

intuitionists to bestow a coherent meaning on their logical constants.

It should perhaps be mentioned that I did not include so-called "second clauses" for '$\rightarrow$' and '$(\forall x)$' in the heuristic interpretation given above. The second clause says of a particular pair of constructions that the first is "a proof of the fact" that the second is a construction which converts each construction for S into a construction for T, or that the second is a construction which associates to each $a \in D$ a construction for Sa. I would certainly agree that since we are dealing in these cases with universal statements we would need to be constructively convinced of their correctness. But just how issues about second clauses should be resolved, would be, I think, a matter of starting with the correct philosophical view and then developing a proper formal analysis on that basis. The idea of understanding the knowledge involved here in terms of fulfillments of intentions should be useful for this purpose.

We have been emphasizing the notion of construction, that is, intuition of mathematical objects, as a process. We should take note, however, that the term "construction" can be used in several different senses. Our discussion of fulfillments of intentions should make it clear what some of these senses are and we would do well to keep them in mind. Sundholm has noted three uses in particular that are important for any theory of constructions.[19] "Construction" may refer to:

(i) process of construction
(ii) object obtained as a result of a process of construction
(iii) the construction-process as object.

These distinctions arise naturally out of our considerations about fulfillments of intentions. We have so far been emphasizing the first and second uses. Fulfillment has been thought of as a process that must be carried out, or as a process in which we may become actively engaged. But we have also been using "fulfillment" in the sense of the output of the process, that is, for the object obtained in carrying out the process. In this chapter we began to say more about how we should understand the notion of "object" in this use in the case of mathematical objects. The third use is the use appropriate to *reflection* on processes of construction. For in acts of reflection the processes can themselves be made objects. In mathematical activity we carry out certain processes but we can also reflect on these, treating the processes themselves as objects. Reflection should be considered a part of mathematical activity, as Brouwer emphasized. As we noted in Chapter 2, acts of reflection are to be viewed as processes which may be fulfilled, partially fulfilled,

frustrated, and so on. In use (iii) we are understanding construction-processes as objects of acts of reflection. We should note that just as in the case of prereflective acts, we may think of fulfillments of acts of reflection as processes of construction [use (i)], or as objects obtained as a result of a construction process [use (ii)]. Acts of reflection are founded acts – they depend on the fact that some process has been carried out and are then directed to this process. The possibility of such reflection is of course an integral part of the phenomenological perspective. Recall that on the phenomenological view we can then reflect on our reflections so that we have hierarchies of acts of reflection.

Let us now make a number of further observations on some specific clauses in our explanation. In the clauses for the quantifiers we shall understand variables as functioning in the way that we suggested in Chapter 3. In particular, the "x" in $(\exists x)$ Sx and $(\forall x)$ Sx should be understood as having a demonstrative element. In veridical intuition "x" expresses the synthetically identified x's of acts in the sequence of intuition in which an object is given. In nonveridical intuition the x's of these acts are not synthetically identified – at some stage x is intuited as an object different from the object intuited at earlier stages. In cases where we intuit different objects, e.g., a chair, a pencil, a pipe, the x's of the acts directed to each object are of course not synthetically identified. This fact can be expressed by use of different variables or use of subscripts. Questions about identity and further matters about the use of variables and constants will be clear when we discuss natural numbers and finite sets.

In the clause for the existential quantifier we must have a construction of an object a, that is, the intention expressed by "a" must be fulfilled. This is in keeping with our comments on intuition *of*. The intention expressed by "Sa" must also be fulfilled, which requires intuition *that*. There is an important observation to be made about the notion of "existence" expressed by the existential quantifier on our interpretation. To say that the intention expressed by $(\exists x)$ Sx is fulfilled is just to say that there is intuition or consciousness of an object as existing. It is to say that there is evidence for belief in the existence of the object, and this is not to be confused with the assertion that the object exists. The existential quantifier, in effect, is in the scope of an epistemic operator. In phenomenology it is not claimed that the existence of an object depends on the possibility of our having an intuition of it. This would be a form of psychologism or of idealism that Husserl attacked even in his later writings. Phenomenology is concerned with an analysis of the awareness of an object as existing, independently of the question whether it

in fact exists or not. This is true whether the objects in question are mathematical or physical. Husserl's various remarks about the phenomenological reduction are attempts to make this point. The adoption of this kind of attitude with respect to both mathematical and physical reality is reminiscent of Gödel's remark that "the question of objective existence of objects of mathematical intuition is an exact replica of the question of the objective existence of the outer world". Having said this, we must course keep in mind that our use of '$(\exists x)$' in mathematics expresses the "meaning of being" appropriate to mathematical objects.

A brief observation about '$(\forall x)$' is also in order. For the intention expressed by $(\forall x)$ Sx to be fulfilled the intention expressed by Sa must be fulfilled for any $a \in D$. If D is finite and quite small we could expect to actually intuit that Sa for each $a \in D$. In that case our evidence for the belief that $(\forall x)$ Sx would be quite good. But when D is sufficiently large we cannot of course actually intuit that Sa for each $a \in D$. In such cases the relevant question is whether it is possible, and so whether Sa is fulfillable for an arbitrary $a \in D$. Our belief that it is possible will be a function of the evidence that we have at a given stage in the intuition. The universally quantified statement is a statement about future intuitions and what is possible in this regard will be determined by the horizon of an act at a given stage in the sequence of intuitions. If, for example, the sequence of future intuitions can be described as lawlike in the sense that the object that will be intuited at each stage is fixed in advance by a rule or algorithm, and there is some insight into the rule, then our evidence for the statement that $(\forall x)$ Sx will again be quite good. We shall claim that it does not follow that there is not intuition that $(\forall x)$ Sx just because we are not actually able to intuit that Sa for each $a \in D$. Our intuition that $(\forall x)$ Sx may be more or less adequate depending on what kind of evidence we have, but it is intuition nonetheless. We have here a case of intuition of a whole founded on intuition of parts. We shall address closely related matters in much more detail in our treatment of mathematical induction in the next chapter.

On our view of logic it is natural why the principle of the excluded middle should fail. For any intention expressed by $S \vee \neg S$ to be fulfilled it must be the case that either the intention expressed by S is fulfilled or the intention expressed by $\neg S$ is fulfilled, i.e., for any intention either it or its negation is fulfilled. In the case of mathematics many (weak) counterexamples to the principle can be given.[20] Consider, for example, the statement

Πn: n is the number of the last decimal of a sequence of ten consecu-

tive numerals 7 in the decimal expansion of π. (We can assume that π possesses a decimal expansion).

At the present time it is not possible to decide whether

$$(\exists n)\ \Pi n \vee \neg (\exists n)\Pi n.$$

That is, we cannot assert that for the intention expressed by $(\exists n)\Pi n$ either it or its negation is fulfilled. This kind of counterexample does not of course show that the statement leads to a contradiction or an absurdity but only that at the present time there is no evidence for asserting it. The claim that there are not weak counterexamples to the principle of the excluded middle amounts to the assumption that we have a method for deciding every statement of mathematics, which is equivalent to supposing that our mathematical knowledge is complete. Martin-Löf, in the paper mentioned above, makes a similar point about the excluded middle: if used in an uninhibited way it would lead to programs that we would not know how to execute. If we think of the principle of the excluded middle as applying not only to mathematical statements but to statements in general then the assumption that there are not weak counterexamples to it implies that the intention expressed by any statement can be decided in intuition. On our view of knowledge this would amount to omniscience.

Considerations of this type suggest, as Brouwer thought, that "mathematics precedes logic". We cannot assume that the principle of the excluded middle or certain other basic logical principles of classical logic (e.g., $\neg\neg S \rightarrow S$) hold generally in mathematics since to do so would have unacceptable epistemological consequences like those just pointed out. Rather, the principle of the excluded middle may hold in some parts of mathematics but not in others depending on the part we are considering. In this sense, just which logical principles hold depends on the mathematics. This is simply a consequence of interpreting logic in such a way that it complies with what we may be said to *know*, as opposed to what we can merely think or intend.

6. IS CLASSICAL MATHEMATICS TO BE REJECTED?

It seems to me that we are not forced by our considerations to actually reject classical logic and mathematics and the view of "standard" semantics that usually goes with these. Repudiation of classical logic would require a much stronger argument than I think is provided anywhere in phenomenology. Thus, I read the import of the logical constants on the above interpretation

differently from some philosophers, perhaps most notably, Michael Dummett. Dummett has attempted to give the kind of strong argument that would be required for the repudiation of classical logic. The argument purports to show that classical logic is to be rejected because it violates the principle that meaning is fully determined by use. In particular, Dummett argues that it is impossible to fully manifest knowledge of the meaning of undecidable mathematical statements if that meaning is supposed to be understood in terms of truth conditions. For we can give neither a proof nor a disproof of any such statement, and we can also not fully manifest our grasp of the meaning of any such statement by simply paraphrasing or restating it, since that would only give the meaning to someone who already understood it. We have to break out of the circle of language, otherwise we would never learn mathematical language in the first place. Dummett thus argues, in effect, that classical logic can be seen to violate some very persuasive views about the manifestability of knowledge and learnability of language in the case of undecidable statements. Hence, classical logic is to be rejected.

I cannot consider the details of this argument here, but it should at least be noted that it depends on a variety of assumptions that can be challenged. It depends, for example, on the rejection of holism about language, and on a view about "complete" or "full" (as distinct from partial) manifestability of knowledge that may be suspect. There are many other points on which the argument could be examined, but for the purposes at hand I would simply like to contrast this view with some ideas suggested by a phenomenological view.

I would like to say that classical logic abstracts from intuitability conditions whereas such conditions are built into our interpretation of logic from the outset. It might even be argued that a theory of intentionality worked out independently of questions concerning evidence and intuition would be compatible with classical logic, mathematics, and "standard" semantics. The objects of classical real analysis, complex analysis, functional analysis, group theory, and such, insofar as they resist constructive treatment, would have to be thought of as "transcendent" in a more radical sense than the objects of number theory or finite set theory, since that would mean that one could not even intuit parts of such objects. Our knowledge of such objects would have to be considered conditional or hypothetical in the sense that we would have at best only indirect evidence for the existence of the objects. This is not to say that we cannot consider logical relationships among propositions about such objects. But if we are concerned with issues about *knowledge* of objects then conditions on fufillability of intentions must be considered. For we are

then concerned not with idealized truth conditions under which the truth of every statement could be presumed to be decided but rather with mathematical "truth within its horizons". The fundamental notion thus becomes that of "fulfillability", not "truth" as understood on a standard account. The moral of the story, as I see it, is not the moral drawn by Dummett. Rather, as I describe it in Kantian language in Chapter 8, it is that we understand far less than we can think or reason about.

The question of the relation of evidence and fulfillability to truth no doubt raises large philosophical issues, as has already been mentioned, but I think that in the approach I develop there is a place for both classical and constructive mathematics. I say more about how I think this could be understood from the point of view of phenomenology in Chapter 8. I think it is fair to say that classical mathematics, where it diverges from constructive mathematics, tends to be quite insensitive to some very important epistemological issues, and certainly to a fine-grained approach to questions about knowledge. There would be a fundamental incompatibility were this not acknowledged to be true of classical mathematics.

CHAPTER 5

NATURAL NUMBERS I

1. INTRODUCTION

We shall now take up the question of what it means to say that there is intuition of the natural number n, where we take n to be any particular number. There are many aspects to this question and we shall not be able to treat all of them fully. The account to be discussed is perhaps best be viewed in terms of what I have called the Kantian strategy; that is, in terms of the question how arithmetical knowledge is possible, how in particular we become aware of number. We shall be considering several cognitive processes which are conditions necessary for the awareness of number, and which therefore begin to provide an answer to the Kantian question. The processes to be considered include collecting, reflecting, abstracting, and comparing. In discussing these processes we shall be investigating what Husserl calls the "origin" of the awareness of number. This chapter and the next will be devoted to developing the account in some detail. Objections will be treated in both chapters and in Chapter 8.

We shall begin by making a few general comments on the notion of number and on what Husserl took to be the purpose of a phenomenological investigation of number. The main point here is that the concept of number is not definable, that it is a primitive concept, and thus that one can only investigate its "origins". Throughout the chapter references will be made to certain views expressed in Husserl's *PA* but to readers of that very early work my development will appear to be different. Husserl of course came to believe that the work on number in *PA* needed to be entirely recast in order to properly reflect his mature conception of phenomenology. The kind of work needed was to some extent undertaken by later writers in the phenomenological tradition like Felix Kaufmann and Aron Gurwitsch.[1] One of the most important needs was of course to expurgate psychologistic confusions in the account of the "origin" of number. This is largely accomplished through keeping in mind the distinction between numbers and their properties on the one hand, and number presentations and their properties on the other. In investigating the latter one avoids psychologism by the Kantian strategy, for it should then be clear that the concern is with epistemology, not with

empirical psychological facts.

Another improvement one sees in the later writers was the effort to place more emphasis on the ordinal conception of number than Husserl had in *PA*. In *PA* the emphasis is on the cardinal concept and one has to read between the lines for views on ordinality. One also sees somewhat more sensitivity to the question of where higher-order aspects of number come into play in the concepts of ordinality and cardinality, and to the set-theoretic development of these concepts.

After some brief discussion of the purpose of a phenomenological investigation of number it is pointed out that arithmetical knowledge must be considered a product of number intentions and intuition. The "horizon" associated with acts directed toward numbers will be discussed and a description of the processes involved in fulfillments of intentions to quite small natural numbers and singular statements about them will be given. The elements of the description of what Husserl calls "collecting" and "authentic counting" will be analyzed in some detail with a concern for both ordinal and cardinal aspects of number. In particular, the temporal structure of the processes and the notion of "synthesis" will be looked into more fully for consciousness of these objects than was done for ordinary perceptual objects in Chapter 3. Another element of the description that will require attention is the role of acts of formal (categorial) abstraction and of reflection in founding the intuition of numbers. Some objections to the notion of abstraction will be considered in this context. Questions about the degree and type of evidence for a mathematican M's beliefs about the existence of quite small numbers and singular statements about them provided by the processes described will be briefly addressed. Questions about large numbers and principles of induction and recursion will be taken up in the next chapter.

2. THE CONCEPT OF NUMBER CANNOT BE EXPLICITLY DEFINED

The purpose of a phenomenological investigation of number is not to define number but rather to offer an account of the "origin" of number. In Husserl's *PA* it is said that the concept of number is not capable of definition and that the only way we can really hope to clarify the concept of number is through an investigation of its origins. In *PA* the notion of "origin" is laden with psychologistic overtones but it is clear from later work that what is sought is an investigation of the processes of the "transcendental ego" that must be involved in construction of the consciousness of number. In *PA* Husserl puts the position as follows:

> One can only define the logically complex. As soon as we come upon ultimate elementary concepts all defining ends. No one can define concepts such as quality, intensity, place, time, and the like... Identity, similarity, increase, whole and part, plurality and unity, and so on, are concepts which are entirely unfit for formal-logical definitions. What we can do in such cases consists only in indicating the concrete phenomena from which they are abstracted, and describing the abstraction processes used... recreating for ourselves the psychological processes required for the formation of the concepts.[2]

Husserl goes on to say that the concept of number is so closely connected with undefinable concepts like that of multitude, unity and plurality that it too is undefinable. An important concept which Husserl does not mention in his list of undefinables is the concept of set or class. In order to see Husserl's point about the undefinability of number one might consider the parallel with any attempt to explicitly define the concept of set. I shall have more to say about this below.

In order to understand the point about the undefinability of number, and the need instead for an analysis of the origin of the concept, it is especially instructive to consider Husserl's objections to Frege's account of number. For as Husserl saw the matter, Frege was trying to do the very thing that could not be done. Frege's work was judged to be fruitless by Husserl, at least for philosophical purposes. Why? The main answer, as I see it, is that it is inadequate as a response to the Kantian question. But let us consider some of the details.

Why would anyone argue that number is undefinable? The argument, as I will develop it, can be viewed in two parts.[3] First, there is an argument based on the "paradox of analysis". Simply put, this is the problem that if one is trying to define the *sense* (in Frege's and Husserl's sense) of an expression of number then the sense of the definiens will either be the same as or different from the sense of the definiendum. If the sense is the same then the definiens simply repeats the definiendum so that one has an obvious circle, in which case the definition is pointless. But if the sense of the definiens is different from that of the definiendum the definiens adds something to the definiendum, in which case the definition is false. Husserl claims that he wants to get at the sense of statements of number, but that this cannot be done through explicit definition. There is thus a problem about sense or meaning for a Fregean account of number, or for other accounts that attempt to give explicit definitions. Frege has not arrived at what is essential about number. But there

is also a problem of *reference* for accounts of number that offer explicit, reductive definitions. Frege does attempt to offer an extensional definition of number and he claims that this is perfectly adequate for mathematical purposes. Husserl of course thinks that the extension of the concept of number is not problematic since no one has any difficulty in applying it to given cases. But an extensional definition, in any case, is not of much value, at least philosophically. I shall not question whether extensional definitions indeed are adequate for mathematical purposes, but there are still other problems with Frege's efforts. The main problem is the multiple reduction problem, for there are infinitely many extensional definitions of the numbers that would satisfy the principles of arithmetic. But then how can Frege claim that he has defined "*the* numbers"? This objection is not to be found in Husserl but one might view the multiple reduction problem as bearing out Husserl's view that the concept of number is really not definable. Perhaps the problem of the indeterminacy of reference of number terms can be seen as a symptom of the effort to explicitly define number. If we take the effort to explicitly define number to be philosophically mistaken then the problem of the indeterminacy of reference of number terms might be seen in an entirely different light.

The problems with Frege's definition of number, from the point of view of phenomenology, are clearly described in the following passage from *PA*:

> If numbers are defined by means of the relation of equivalence (1-to-1 correlation), then every assertion about number is always directed – instead of at the concrete set at hand – at the relationship of the set to other sets. To attribute a definite number to this set means classifying it with a definite group of equivalent sets; but this is not at all the sense of an assertion of number. Let us consider an example. Do we call a set of nuts lying before us four, because it belongs to a certain class of infinitely many sets which can be put into mutual unique correspondence? Of course no one ever had such thoughts when doing that, and we can scarcely find any practical motives at all for us to be interested in the same.[4]

On this view one of the serious problems with Frege's account, and this generalizes to other reductive accounts, is that number, on Frege's definition, is not what is "before the mind" in awareness of number. To see the point of this argument think of substitutivity *salva veritate* in the context of epistemic operators. It does not, for example, follow from the fact that M believes that there are three coins on the table that M has any beliefs about the class of all triples. One might very well have the former belief, and would usually have

it, without having the latter belief. Examples of this kind can be multiplied indefinitely and applied to any reductive definitions of number.

We might press the point farther, given what we have said about intentionality, by noting that Frege's definition does not describe the process we actually undertake to determine number. On a constructivist view of definition, a view entailed by concern for fulfillability of intentions, that is exactly what a definition should do. Furthermore, if we look at Frege's definition from this point of view then it is not clear that it even gives a procedure that a person *could* undertake, unless it is understood as not assuming an actual infinite, or as not making other non-constructive assumptions. Of course Frege did not think of definitions of number as descriptions of processes for determining or finding number, but insofar as he did not his view fails as an answer to the Kantian question of how awareness of number is possible. It could be argued that for these reasons Frege's effort fails in some important respects on questions concerning the epistemology of mathematics. It seems to me that Husserl would surely criticize Frege for not having an epistemological account of arithmetic. Logicism would certainly fall short of providing an adequate account. Husserl, like Frege, recognizes a sense/reference distinction, but for Husserl numbers, while taken to be objects, could not be considered "logical" objects.

3. THE ORIGIN OF THE CONCEPT OF NUMBER

If the concept of number is not explicitly definable, if it is a primitive concept, then how is it to be understood? One might ask this question of the primitve notions of any mathematical theory. Husserl's answer is that the concept of number is to be understood through an analysis of its origins. On this kind of analysis it is assumed that there is awareness or experience (and then, in particular, intuition) of number and the problem is then to explain how this is possible.

Before we go much farther we should examine the assumption that there is experience or awareness of number in order to be clear about how it should be understood. I shall argue that there is nothing at all mysterious about the awareness of number. Here is an example of what I mean. Suppose we are sitting in a classroom and I ask you to consider a certain group of objects on a table, let us say a cup, a pen, and a book. If you do consider the group of objects then clearly you are aware of something that you were not aware of before I initiated the example. There is a different act of awareness, there is something that you are aware of now that you were not aware of a moment

ago. Now suppose I ask you to determine the number of objects in the group. You can probably do this in one glance since it is a very small number. But suppose I had picked a larger group. Then you are going to have to undertake some process in order to tell me what the number is. This is again a case where something about your awareness changes between the stage of your experience at which you simply saw the group of objects and the stage at which you determined the number of objects in the group. First you simply saw, for example, a group of cups and then you saw the cups *as nine cups*. There is something analogous to a Gestalt shift in your awareness in these cases, for it is the same underlying data that you are looking at, but you are aware of the data in different ways. Husserl would say that you have the same hyletic data in these cases but that you are interpreting it differently, that different acts are involved. If you have followed this example I think you would be hard-pressed to deny that you are aware of the number of objects in the group, or that you *know* how many objects are in the group. I might then go on to ask you, for example, whether there are as many objects in the group as there are strokes written here – | | | | | | | | | | | – and you would carry out another process to make that determination.

Now suppose we work through several examples like this in order to determine what they all have in common insofar as your awareness of the number in each case is concerned. What they would have in common is something about the process you carried out to determine the number, not anything about the particular objects that you happened to count or collect in each case. Hence, the view goes, what is relevant to the awareness of number is not anything about the particular objects numbered, but rather the structure of the cognitive processes involved in making determinations of number, a structure which is evidently isomorphic across the cases. The structure is also evidently isomorphic across persons who make number determinations, a topic we shall discuss later.

In considering these examples we are already investigating the origins of number, in Husserl's sense. We are considering the simplest examples of experiences out of which awareness of number arises; that is, we are considering the processes which produce M's beliefs that S, where S is a very elementary singular satement about numbers. The notion of the origin of an idea or concept here is of course not to be understood as a classical empiricist would understand it. That would probably lead to psychologism about the concept of number. Consider, for example, Hume's view about the origins of concepts: all ideas or concepts are derived from (sensory) impressions. Husserl agrees that the awareness of number is in a certain sense "founded"

on sense experience, but claims that, as Kant emphasizes, knowledge is not exhaustively determined by sense experience. Certain cognitive structures and processes are involved in knowledge, these are contributed *a priori* by the mind, and any analysis of the origins of knowledge must seek to determine these structures and processes. In particular, an understanding of these structures and processes will help us to determine how we could possibly have knowledge about objects like numbers and finite sets, when these objects are thought to be abstract.

In *PA* Husserl's genetic investigation of number thus begins with the familiar, everyday phenomenon of experiencing groups of ordinary perceptual objects as groups or "multitudes" (Vielheiten). The notion of "multitude" is used quite naively by Husserl in *PA*, especially in light of Frege's objections to understanding the concept of number in terms of the notion of multitude. We shall have occasion to discuss Frege's objections below, especially his objections about the numbers 0 and 1, and about the necessity of distinguishing what we would now think of as mereological sums from sets (or extensions of concepts).[5] Our treatment of questions specifically concerning constitution of the consciousness of finite sets will be mostly developed in Chapter 8. The question of intuition of finite sets is of interest in its own right in connection with the claim of logicians like Gödel and Wang that there is something like intuition of sets. (Of course Gödel and Wang are speaking of transfinite sets but the claim is surely most plausible in the context of finite set theory.)

Let us simply note for now that for Husserl in *PA* number is viewed as the genus of which particular numbers are said to be the species. We are aware of numbers as determinate multitudes. In *PA* the concept of number is distinguished from the concept of multitude in general insofar as number always involves a "precisely determinate *how many.*" Particular numbers are said to arise through specification of the concept of multitude. When we enter a room full of people, for example, we may be aware that there are a number of people. We could then set about determining how many people there are in the room. Or similarly, if we experience | | | | | | | | | as a number of strokes we could then go on to determine its number. Much of the analysis in *PA* is then concerned with the processes necessary for determining the number of objects in a multitude. We shall therefore start by considering the processes involved in determining how many x's there are such that Sx, where we think of the objects involved as ordinary perceptual objects. Later we shall discuss what is involved in more general applications in which the objects can be any objects at all. It is by way of the processes involved that there is intuition of

particular natural numbers.

4. INTUITION OF NATURAL NUMBERS

Arithmetical knowledge is to be viewed as a product of intentions and intuition just like any other kind of knowledge. Intentions directed to particular numbers refer not only to objects that may be thought to exist independently of us and our constructions but also to certain kinds of expectations about the course of our experience with respect to these objects. We can view an empty number intention as an expectation that a certain number can be found. As such an intention may be indefinite, as when we expect e.g., that "there are a number of objects", or definite, as when we expect, e.g., that "there are five objects". Associated with the number intention will be a horizon of possible experience in which the intention to the number would come to be fulfilled, partially fulfilled, frustrated, and so on. In the case of intentions to particular numbers the horizon is quite well-determined. The process for finding an object emptily intended is in fact lawlike. We might describe the process in which intentions to particular numbers are fulfilled in different ways but we would want to know that the descriptions were isomorphic to the description that will now be given. On the basis of our remarks on quantifiers the description might at first be viewed as part of a description of the fulfillment of the intentions expressed by the numerically definite quantifier $(\exists_n x)$ Sx: "there are n x's such that Sx". To say, for example, that there are two x's such that Sx is to say that

$$(\exists x_1)(\exists x_2)(Sx_1 \wedge Sx_2 \wedge x_1 \neq x_2 \wedge$$
$$(\forall y)(Sy \rightarrow (y = x_1 \vee y = x_2))).$$

The use of identity and negation permits saying in conformity with the notation of Chapter 3 that the x's, $x_1, \ldots, x_n$, are not synthetically identified. Out of some considerations about ordinals and the relation of intuition of numbers as ordinals to the intuition of numbers as cardinals we shall first pass through the ordinal quantifier, $(\exists_{nth} x)$ Sx: "there is an nth x (relative to an ordering) such that Sx." We shall take the ordinal quantifier to be relative to an ordering. As we proceed we shall be discussing conditions for quantification on the numerical variable n itself.

We can view the intuition of natural numbers as founded on the structure or form of sequences of our perceptions by abstracting from the qualitative differences in the mental acts of which they are composed. This is what

emerges from consideration of cases like those given above. One comes to see that nothing about the particular objects collected or counted on a given occassion is relevant to their number. What is isomorphic across the cases is a certain structure involved in determining the number. We now describe this structure. The idea of iteration or succession in time is of course an essential feature of the process of intuition according to the phenomenological description of this process. We can describe the fundamental process involved in intuiting numbers in the following way: Natural numbers are constructed at successive stages in time. At a certain stage we construct a unit, at a later stage we construct another unit, and then we look at this pair as an object. We could then at a still later stage construct another unit, taking the pair previously formed as a term of a new pair, which gives another object. This process could be iterated as often as we like, as represented by the following diagram:

$$\begin{array}{l} S_1 \quad S_2 \quad S_3 \\ (((\,|\,)\ |\,)\ |\,)\ldots \end{array}$$

In continuing the construction we are thus simply iterating the process (operation) of constructing a unit, i.e., adding a unit at each stage where we think of this as done at successive stages in time. The idea of constructing a unit in our description should be understood in terms of the determinable x component of an act. At each stage in a construction we perform an act of constructing a unit, or a "something", an "x". The units are simply placeholders in a construction and, by virtue of their indexicality, they could be anything at all. At each later stage in the process the structure that was given at the immediately preceding stage is retained as it was originally given. So as we continue the construction each later stage has built into it, so to speak, exactly what was given at all of the earlier stages. The process for fulfilling number intentions of the type we are considering is thus quite simple and well-determined: it might be viewed as amounting to iteration of a successor operation. Because of its simple lawlike character it is a much less complicated process than is involved in most of our attempts to obtain knowledge. This fact has of course impressed a number of mathematicians and philosophers.

We should discuss the elements of this description in some detail. We might first note that the description bears some resemblance to Brouwer's description of the "first act of intuitionism" on the basis of which finite ordinals are to be regarded as constructible.[6] As a general observation we should note however that there have been a variety of versions of the view of

numbers as multitudes of units obtained by abstraction ever since the time of Euclid's definition of number. Weierstrass held such a view, for example, and so did Cantor. Cantor's views in fact seem quite close in certain respects to those of Husserl in *PA*.[7] What Brouwer emphasized and what we need to keep in mind is that in our description we are speaking about the structure of the mental processes involved in the construction of natural numbers, i.e., about how we come to be aware of numbers. We are giving an account of numbers as objects obtained as a result of a construction process.

5. ORDINALS

Awareness of natural numbers as ordinals or cardinals will arise depending on whether we regard the units of the construction as ordered or not. Awareness of an ordering does not of course result only from our own ability to actively construct units. In hearing the peals of a bell, for example, we are aware of their ordering in time and perhaps of their ordinal number, but the peals are already given to us. Similar remarks apply to other space-time objects or events that are given to us in some order. What I shall be interested in is the structure of consciousness which makes possible the awareness of order in any case where there is such an awareness. In our case of constructing units an ordering of units is imposed by virtue of their being constructed at successive stages in time. We might say that time provides a universal source of models for number terms. It is not as if, in constructing units, we merely construct "something and something and something" without any ordering, but rather we must construct the units in a definite order, one after another. We would normally describe this order by saying that certain units come "before" or "earlier than", or "after" or "later than" other units, as we continue the construction. This ordering in part underlies the intuition of "<" and ">" relations among numbers. To express or name the order that results from the temporal structure of consciousness we would say that we construct a *first* unit, then a *second* unit, then a *third* unit, and so on. These expressions or names merely indicate the relation in which the units stand to one another as we continue the construction. At the nth stage of the construction we construct the nth unit. The construction is to be viewed as a process that must be carried out if intentions to ordinals are to be fulfilled. For then we do not merely expect that we can find a first unit, a second unit, and so on, but we actually perform the acts in which the number is given.

The temporal structure of consciousness which makes possible the construction of ordinal numbers has been described in some detail by

Husserl, even though he did not specifically consider it in relation to this problem.[8] Before we consider some details of Husserl's description we might note than an earlier but somewhat similar view of the role of the temporal structure of consciousness in the constitution of the consciousness of number can be found in Kant's comments on number. In his discussion of the "synthesis of recognition in a concept" in the first edition transcendental deduction, for example, Kant points out that

> If, in counting, I forget that the units, which now stand before me (die mir jetzt vor Sinnen schweben), have been added to one another in succession, I should never know that a total is being produced through this successive addition of unit to unit, and so would remain ignorant of the number. For the concept of number is nothing but the consciousness of this unity of synthesis.[9]

At a later point in the *Critique* Kant argues that "The pure schema of magnitude (*quantitas*), as a concept of the understanding, is *number*, a representation which comprises the successive addition of homogeneous units." Kant concludes there that "Number is therefore simply the unity of the synthesis of the manifold of a homogeneous intuition in general, a unity due to my generating time itself in the apprehension of the intuition".[10] On the kind of account we are giving it would be better to say that the awareness of number, not that number itself, results from the unity of such a synthesis. Of course there are indications elsewhere in Kant's writings that he would take a similar view.

We should now consider some of the features of the temporal structure as these are analyzed by Husserl. Suppose we consider a simple construction in which we construct five units. In this construction there will be one act for each unit that we construct and these acts will be performed in succession at, say, times t_1, t_2, t_3, t_4, t_5. Now an important feature of this process of construction for the awareness of ordinal numbers is that at each later stage what was constructed at earlier stages is *retained* in an appropriately modified way. Husserl describes this process of retention in some detail. At t_4, for example, what was constructed at t_3, would be retained as "just past", what was constructed at t_2 as further in the past, and what was seen at t_1, as even further in the past. This type of retention is a very basic feature of consciousness which is involved in the constitution of any kind of object. It should be noted that it is not to be confused with an *act* of remembering. It is not an act that is performed but is rather a "passive" feature associated with any act that has to do with our immediate awareness of an object.

We can use a diagram to represent the details of this structure in the case of our example. We can suppose that in constructing a unit at each stage we make an inscription on a piece of paper, as I did earlier:

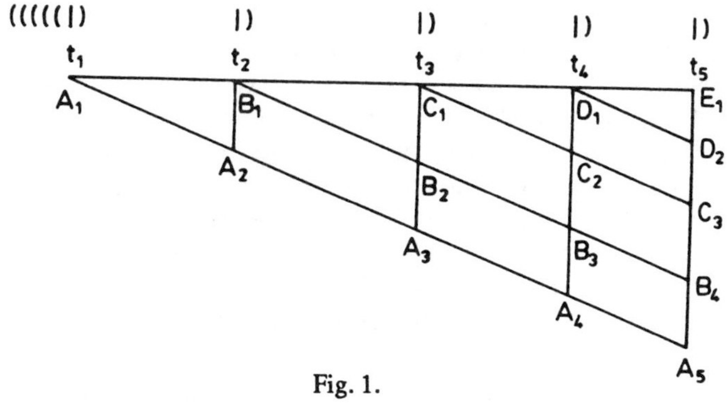

Fig. 1.

At t_1 we construct the "first" unit, at t_2 we construct the "second" unit, and so on. The horizontal line A_1E_1 represents the sequence of acts. The diagonally descending lines A_1A_5, B_1B_4, and so on represent sequences of retentions. The sequence of retentions represented by the descending line A_1A_5 (not including point A_1) are retentional modifications of what was constructed at t_1, the sequence of retentions represented by the descending line B_1B_4 (not including point B_1) are retentional modifications of what was constructed at t_2, and so on. Now we can represent the degree of "pastness" of a retention by the vertical distance of the retention from A_1E_1. Numerical subscripts have been used to facilitate the idea that what was constructed at earlier stages is retained in "an appropriately modified way". This gives the sense of our remark that at t_4 what was constructed at t_3 is retained as "just past", while what was constructed at t_2 is retained as further in the past, and so on. The parentheses around the inscriptions can be taken to represent what we have constructed at a given stage when we look down the vertical lines in the diagram. We could just as well use standard notation for ordered n-tuples if we wished to convey the ordering.

It should be clear how this kind of structure is a condition necessary for the awareness of position of units in the construction in relation to one another. If, for example, what was constructed at t_1 or t_2 was not retained (i.e., was forgotten), at t_3 there could not be a correct awareness of the ordinal number

of units at t_3. There would also not be correct awareness of the ordinal position of the unit at t_3 if only t_1 were retained, or only t_2, or if t_1 and t_2 were not retained in an appropriately modified way (i.e., if the retention were distorted): what is constructed at t_3 could not be present in the same manner as what was constructed at t_2.

In the case of larger constructions it will of course happen that at later stages what was constructed at the earliest stages will not be retained as something of which we are still immediately aware. However, because what was constructed at each earlier stage is at least retained in the act which is its immediate successor, it is possible to construct larger ordinals. For suppose we are at some late stage in the construction which occurs at t_k. Then clearly at t_k there will at least be retentional awareness of what was constructed at t_{k-1}, although not everything will be immediately present in the act that occurred at t_{k-1}. But surely what was constructed at t_{k-2} will be retentionally retained in the act at t_{k-1}, although not everything will be immediately present in the act that occurred at t_{k-2}, and so on until we reach the first stage of the construction.

What is important about this picture is simply that the information we have in an act at some stage about the ordinal position of the unit constructed at that stage is carried into the act which is its immediate successor. This feature of the way that acts in the construction interpenetrate makes possible the intuition of large ordinals. The evidence that we have for large ordinals, however, will differ in degree from the evidence that we have for smaller ordinals.

The process discussed so far would be necessary for intuitive awareness of ordinal numbers. In the case of small numbers the process can actually be carried out. Certain higher-order features normally associated with the ordinal conception of number, in particular, for identity conditions on finite ordinals, will be considered in the next section.

A point that I shall only mention at this time is that a correct understanding of the temporal process involved in the intuition of ordinal numbers should make it possible to avoid Frege's charge that in explanations of the sort we are giving it is impossible to reconcile the sameness of the units of which we speak with their distinguishability.[11] What Frege had in mind was that explanations of number that appealed to constructions involving "units" or "mere somethings" could not provide an account of how units could be distinct from one another since in obtaining them by "abstraction" we are to think of them as divested of all properties that would enable us to make such a distinction. I shall have more to say about this idea of constructing units or

"somethings", but it should be clear from what we have said that this is really not a difficulty at all. For in constructing units in the process we have described it is clear that at the very least the units will be distinguishable by virtue of the degree of modification in our retentional awareness of them at a given stage in their construction. Since we shall understand the construction of "units" in terms of the determinable x of each act, we can say that each act has its x, and that in a correct determination of number these x's will not be synthetically identified. Of course it might turn out, as in a mistake in counting, that the x's (or some of them) are "synthetically identified", so that there are fewer objects than there are acts. But even here there is distinctness of units by virtue of the fact that the x's are x's of acts carried out in temporal succession. This factor will also play a role in the intuition of numbers as cardinals. Frege's point was mainly directed to the cardinal conception of number.

6. ORDINALS AND CARDINALS

We have so far discussed the process which is fundamental for intuiting the nth unit in a construction, and hence for the ordinal conception of number. Much remains to be said about this process, especially about the idea of "construction of a unit" and about what was referred to earlier as "abstracting from qualitative differences" in the mental acts of which a construction is composed. However, we shall first discuss how this process is also involved in the intuition of the number n, and hence in the cardinal conception of number.

From the point of view of the constitution of the consciousness of number I would like to say that the cardinal conception of number is founded on the ordinal conception, or that the intuition of numbers as cardinals is founded on the intuition of numbers as ordinals.[12] The reason for this is that for larger constructions we cannot determine the number of units in one act but must successively run through the units to determine number. I shall say more about this in a moment. The basic idea is that to be conscious of a cardinal number n it is necessary to abstract from the ordering of units in a construction that would be constituted by virtue of successively running through the units. I do not think the position I outline necessarily conflicts with other views, like Russell's, that take the cardinal conception of number to be more basic or fundamental since I am only speaking about the matter from the point of view of the *mental processes* that must be involved.

In the description in section 5 it was said that at a certain time we construct

a unit, and at a later time we construct another unit, so that we have a "first" unit, and a "second" unit. *Then* we look at or regard this pair as an object, as *two* units. After constructing a "third" unit we take the pair previously formed as a term of a new pair, and *then* look at the resulting pair as an object, as *three* units. In general, at the nth stage in the construction where we reach the nth unit we shall, in looking at the new pair formed at that stage as an object, have n units. In this process intentions to numbers as cardinals are fulfilled. For then we do not merely expect that "there is one unit", that "there are two units", and so on, but we actually perform the acts in which the number is given. We have a verification of the existence of the number.

We should now ask what is involved in looking at the pair as an object at any stage, so that we have a construction of the number n. What we are doing obviously amounts to taking the last place corresponding to the last stage in the construction – the nth – to characterize the result of this process when we view it as a *unity* or *collection* of n items. The expression for the place or position of the last unit is looked at as the expression for the unity or collection which is constituted at that stage, hence as n. Once again we can describe the structure of consciousness which makes it possible to construct these objects. We can consider the same process of retention described in section 5 and represented in Figure 1 except that now we shall focus on a different aspect of this process. If we consider a particular stage in the construction and look along the vertical line in our representation of that stage we can say that there is a *unified* or *synthesized* consciousness of what was constructed at earlier stages by virtue of the fact that all of the previous stages are retained at that stage. We might say that they are "contained" (as parts) in the act at a given stage. Suppose, for example, that we are at t_4. Then what was constructed at t_1, and t_2 and t_3 is retained in our act at t_4. At t_4 one term of the "pair" contains what has been constructed up to (and including) t_3 and the other term is simply the unit constructed at t_4.

In order to bring out more forcefully what is involved in this structure of consciousness let us suppose that we are presented with a string of inscriptions:

|||||

If we see this string as a number of strokes we can go on to solve the problem of how many strokes there are. To solve this problem is a way of fulfilling the intention that "there are a number of strokes". As we run through the string there will be one act directed to each stroke and the process of running through it will have the retentional structure already described in section 5. In

addition, according to Husserl's analysis there will be a *protentional* structure. What this means is that when I have reached the third stroke, for example, I will be aware that there are further strokes to be taken into account. I anticipate or expect to run through some strokes and this expectation is part of my act at t_3.[13] Note that once I see the string as a number of strokes and before I begin to determine how many strokes there are I have what might be called "pure" protentional awareness, or pure expectation that there are a number of strokes. It is only once I begin to run through the strokes that this empty intention comes to be fulfilled or realized. As I run through the strokes I experience strokes that I will take account of in the future as being less and less future, finally becoming accounted for, and then receding into the past. Just as in the case of retention, there will be degrees of modification in our protentions. We can present the protentional aspect of consciousness of the strokes by simply adding to our earlier diagram:

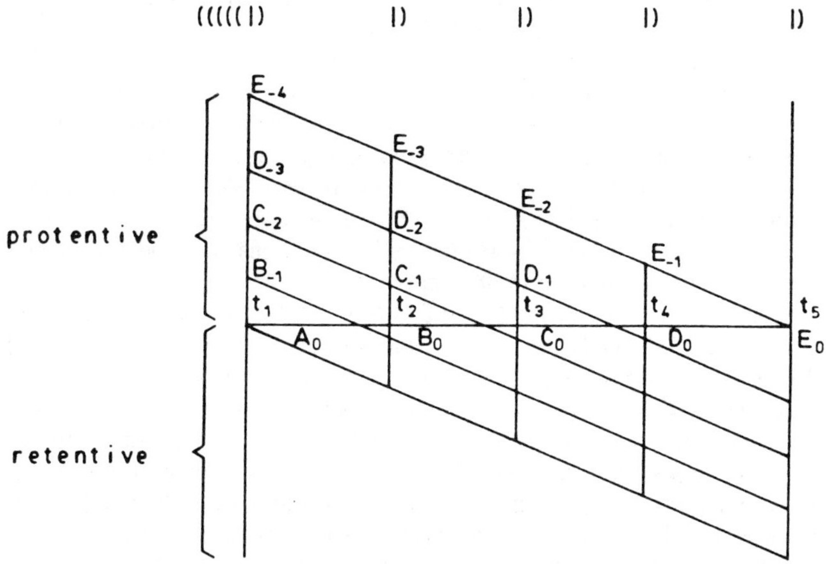

Fig. 2.

The retentional structure is the same as before. Now we simply note that at t_1 we see the first stroke, and with varying degrees of anticipation expect to take account of the other strokes as well. As we run through the strokes we

eventually take account of each stroke, and what was seen at previous stages is retained.

This kind of structure underlies perceptual awareness just as it underlies consciousness of numbers. Husserl points out in *PITC* that all acts are characterized by retentive-primal impression-protentive structure. As we said in Chapter 3, this structure is to be considered part of the immediate horizon associated with any act. The terms "protention" and "retention" are taken to designate those features of an act that are responsible for our awareness of the immediate past and the immediate future at any given point in our experience. We shall have more to say about the function of protention in the construction process for natural numbers in our discussion of mathematical induction.

The point to be made at the moment is quite simple. Namely, that at any stage in running through some objects there is an important difference between what has been seen and what has not yet been taken into account but is only anticipated. At t_3 it is clear that what was seen at t_1 and t_2 is retained in our act at t_3. But what has not yet been taken into account is a stroke or strokes of which we may be aware, i.e., they are protended, but it (they) are not yet unified in a single act with those strokes that have been accounted for. On Husserl's analysis this marks a fundamental difference in our consciousness of the items in this group: Some items have been explicitly seen and some items have not. The strokes that have been seen when we are at t_3 are unified or synthesized into a whole in *one* act of consciousness. But the other strokes are not yet unified in one act (see Fig. 1). They will become so as we continue to run through the group and this will happen automatically or, as Husserl says, passively, by virtue of the act-structure involved. In our earlier discussion we focused on the fact that in running through a group there were *many* acts – one act for each item. Now we focus rather on the fact that a particular stage we have *one* act which contains what was seen at all of the previous stages. It is this structure of consciousness which is necessary for viewing the result of a construction at the nth stage as a unity or collection of n items. I shall have more to say about this process of unifying or collecting in discussing the intuition of sets, where it may be understood as the process upon which intuition of sets is founded.

More must be said at the moment about the fact that the result of a construction at the nth stage may be viewed as a collection of n items. What is important for seeing that "n" will be an expression for a cardinal number is that the *order* in which we construct units will not make any difference to the result obtained at a certain stage in a construction.[14] In order to make this

clear let us consider again the case where we are already presented with a string of strokes,

| | | | | | | | |

where we view this as a representation or instance of a construction. What is involved in the intuition of cardinal numbers is seeing that we can run through this group in any order and get the same result. In each case we would carry out a different process by running through the group in a different order. When the number n of items is quite small we can actually verify that there will be n! different order arrangements. When will any pair of such processes give the same number? It would be very natural intuitively to say that we know they give the same number when and only when we can bring the units constructed in each process into 1-to-1 correspondence. Attempting to bring the units into 1-to-1 correspondence is of course to be viewed as a mental process that can be fulfilled, partially fulfilled, or frustrated. In the case of small numbers we can again actually complete the process of pairing up units from the different constructions. We should note that this is a place where the use of (symbolic) representations of units becomes indispensable for avoiding error. For in order to bring the units of different constructions into 1-to-1 correspondence it will typically be necessary to use representations to keep track of earlier constructions.

We might say, therefore, that in cases where we consider small numbers the intuition of cardinal numbers involves an "abstraction" from order. The acts of abstraction involved here are not any kind of mysterious process for they are something we can actually carry out by running through this group in different possible orders and then bringing units from each construction into 1-to-1 correspondence. The abstraction which this process constitutes is founded on the different possible order arrangements. The abstraction picks out what is common to the different order arrangements. As a consequence of the process we can see that order does not matter to the fact that the last stroke in each case is the ninth. A finite cardinal number may therefore be viewed as what is *invariant* with respect to change of order in the process of running through the items. A finite cardinal number might be described as the "fixed point" of all variations of running through the group.[15] From the point of view of a development of number theory it might be said that we are defining (in a constructive sense) cardinals in terms of ordinals.

On this view we should not think of cardinals as numbers in which there is no ordering among units, but rather as numbers in which there could be any ordering among the items. We can therefore say that while the cardinal

number of a collection is independent of any particular order in which it is constructed, it is not independent of all such orders, so that order arrangement is genetically prior to the intuition of cardinals. This gives the sense in which we say that the intuition of cardinals is founded on the intuition of ordinals. The act of abstraction involved here just amounts to seeing the invariance with respect to change of order as one of which the different orders are varieties.

If we can see that the order does not matter to the end result of running though a group then clearly the items of the collection can no longer be thought of as holders of specific places. So now we see them as units or "mere somethings", but they also lose their specificity of position. They become holders of some place or other, but not of a specific place as in the case of ordinals. We can thus think of them as arithmetical units in the strict sense of the word. For first we said that we have abstraction from all qualitative aspects of the mental acts of which a construction is composed and now in discussing the constitution of the consciousness of cardinals we also have an abstraction from specificity of place. This might be compared to Cantor's conception of the abstraction involved in the cardinal conception of number as embodied in his $\overline{\overline{A}}$ notation.

It should be clear why we said that the intuition of cardinal numbers is founded on the intuition of ordinal numbers. For in carrying out a construction we shall always be performing it in *some* order.[16] It is only once this is done that it is possible to abstract from the order. The act of abstraction is founded on the process of carrying out the constructions and may be fulfilled, frustrated, and so on. It gives intuition of numbers as cardinals. As a result of our analysis we might say that order is always an implicit (but not explicit) part of the intuition of cardinal numbers. This answers Frege's objection mentioned earlier.

Since we have appealed to the notion of 1-to-1 correspondences in the description of the intuition of natural numbers at this point we should make a few more comments about the matter here. In *PA* Husserl discussed 1-to-1 correspondences in his own account of number primarily in connection with the process of "comparing" multitudes.[17] The process of comparing was said to be essential for "precise determination" of the number of units in a multitude because by virtue of it the relations of equality and inequality for numbers are determined. In being able to determine equality or inequality of multitudes we have conditions for identification and individuation of numbers. In *PA* Husserl says, speaking of the cardinal concept, that "The multitude relations of equal, more and fewer condition the origin of the

number concepts in an essential way... In order for us to move beyond the concept of a 'multitude of units' and form the series of numbers two, three, etc., we must classify multitudes of units, and this requires judgments about equality and inequality... "[18] The process of comparing multitudes or constructions amounts to attempting to bring their units into 1-to-1 correspondence. In *PA* Husserl noted that

> If we think of the act of comparison as a successive process, as it usually is, then
>
> 1) to each element of the group with which we begin there must correspond a like one (ein gleiches) in the other group, and indeed a new one at each step;
>
> 2) in the second group no elements must be left over... Only if these conditions are met can we say that the two multitudes as such are equal... .[19]

(For ordinals we need only note that the 1-to-1 correspondence must be order preserving.) As Husserl sees it, comparing is involved in determining where numbers stand in a sequence in relation to one another, thus making it possible to fully individuate numbers. For it is through this process that we know, for example, that a construction with nine units and a construction with eight units both have more units than a construction with seven units, that a construction with nine units has more units than a construction with eight units, and so on.

In appealing to the notion of 1-to-1 correspondences in our description of the intuition of natural numbers at this point it might be asked whether we are committed to *their existence*. Do we need to take the description as saying that such correspondences *exist* in order for there to be intuition of natural numbers? The correspondences needed will be finite and so we can view them as finite sets. I shall argue in Chapter 7 that it should be possible on a phenomenological view to understand finite sets as objects of intuition. No set-theoretic principles beyond the principles discussed in Chapter 7 would be needed but there would be an ascent in rank involved in moving from "units" to 1-to-1 correspondence between units.

7. CONSTRUCTING UNITS AND THE ROLE OF REFLECTION AND ABSTRACTION

Several further elements of the basic description of the construction process

for natural numbers given earlier should be discussed: (1) the idea of "construction of a unit" and (2) the idea that construction of natural numbers is founded on the form of sequences of perceptions by abstraction from qualitative differences in mental acts of which a construction is composed.

The idea of constructing a unit in our description should be understood in terms of the determinable x component of an act. At each stage in a construction we perform an act of constructing a unit, or a "something", an "x". The phenomenon from which the intuition of number is built up is that of determining that there is an nth x such that Sx or, in abstracting from order, of determining how many x's there are such that Sx, where the x's are objects of straightforward perception. The objects could be, for example, eggs, chairs, people, and so on. In these cases we successively take account of some objects that are *already given* in a sequence of acts. In this case of constructing or intuiting an x the x is given in advance. As we said in Chapter 2, straightforward perception provides the concrete, immediate, and non-reflective basis for all of our experience and intuition of mathematical objects must be constituted from this basis. Acts of mathematical intuition, as well as other categorial acts, must be ultimately founded on acts of sense-perception. This can be viewed as an example in the case of our knowledge of numbers of the older Kantian theme that all knowledge begins with sense experience but does not end there. For we are saying that the intuition of natural numbers is founded on straightforward perception but not that it is of straightforward perceptual objects. It is of numbers and this requires founded acts of reflection and abstraction. One might also think of traditional geometry in this context. For on our view geometric intuition would consist of carrying out construction processes that would yield certain construction-objects. In the case of traditional geometry the construction processes are relatively "down to earth". In the most elementary cases they could be represented by concrete ruler and compass constructions. But even here the construction processes must be viewed as being about abstract objects. The concrete ruler and compass constructions could at best be taken to represent the objects of geometric intuition. Founded acts of abstraction and reflection would have to be involved in geometric intuition.

In the case of the concept of number it is clear that this concept does not apply only to objects of straightforward perception. It is more general. Even in the case of objects of straightforward perception we suppose that the concept of number applies to *any* such objects and that the particular properties of the objects under consideration, such as being of a certain color, shape, and so on, are irrelevant to the knowledge of their number. The idea of

construction of a unit in our description must therefore be understood in a very general way. The units are simply placeholders in a construction. By virtue of indexicality of the acts involved the units could be anything at all. That is as it should be since it should be true that anything that can be conceived as an object can be numbered. The notion of number should be applicable to anything at all. In Husserl's work this understanding of the idea of "units" or "somethings" goes hand in hand with recognition of the *formality* of the concept of number. We can say that in applications of the concept of number the relevant intentional context is provided, so that the x's have content. We number certain definite but not necessarily concrete objects. These objects might be perceptual objects, imaginary objects, mathematical objects or whatever. As we said in Chapter 3, it is the sequences of acts and their properties that are to be considered in phenomenology. In Chapter 3 we of course already discussed the structure or form of sequences of perceptual acts in some detail and noted that the structural feature of acts that was responsible for reference to objects was the so-called "determinable x".[20]

In *PA* Husserl had already given the following description of what was necessary for awareness of natural numbers: "Disregarding the particular features of the synthesized individual contents, one considers and retains each content only insofar as it is 'something' or 'one', and thus, with reference to their collective connection, one acquires the multitude-form pertaining to the given multitude... "[21] As we said above, the intuition of natural numbers is founded on the structure or form of the sequences of our perceptions by abstracting from the qualitative content of the mental acts of which they are composed. The abstraction here involves qualitative aspects of the mental acts but not order. Obviously then acts of formal abstraction and reflection must be involved in intuition of natural numbers. These acts are founded on the process of successively running through some ordinary perceptual objects. Once such underlying acts have been performed it must be possible to reflect on the structure or form of what was done. In attempting to determine the form of what was done it is not, so to speak, the objects from which we are abstracting but rather the qualitative aspects involved in the sequence of acts in which the objects are given. One then sees that the same process can be applied to different objects – to objects of any type. Were there not something like acts of collecting, reflecting and abstracting there would not be intuition of number. This should be obvious in the case of collecting. But the kind of shift involved in first seeing some objects as a group, or in seeing a group as related to another group, or as having a certain

number, or in any of the other ways in which one might change one's awareness of the same objects, also requires reflection. Abstraction, properly understood, is also a condition necessary for intuition of number.

In Chapter 2 we said that acts of formal abstraction are to be viewed as attempts to determine the form or structure of what is given. Acts of formal abstraction and of reflection have the same structure as any intentional acts and can therefore be fulfilled, frustrated, and so on. Because they can be characterized by the difference between intention and intuition they must be viewed as processes that produce certain beliefs – in the case of acts of formal abstraction, beliefs about the form of what is given. What is being claimed about acts of formal abstraction here involves nothing especially mysterious. The sense in which talk of acts of formal abstraction should be understood is simply illustrated by examples from elementary logic courses where we say that a person has or has not been able to determine the structure of a given sentence or argument, or from elementary mathematics courses where a person sees or does not see the structure of the particular phenomenon under consideration. Obviously beliefs about the structure of what is given in such contexts sometimes require correction. If we did not see the structure of what is given in the way that we do in these contexts we would not have mathematics as we know it, and it could be argued that in mathematics that is being created the structure of the phenomenon under consideration is often just what is not being seen, or not being seen clearly enough. Perhaps Bernays had this kind of conception of structure in mind many years ago when he remarked that "mathematics can be regarded as the theoretical phenomenology of structures". Bernays suggested that "what contrasts phenomenologically with the qualitative is not the quantitative, as taught by traditional philosophy, but the structural, i.e., the forms of being aside and after, and of being composite, etc., with all the concepts and laws that relate to them".[22] It seems that, properly understood, the claim at least *that* we have such acts of formal abstraction could not be doubted.

For the proper interpretation of our talk about the structure of the sequence here, or about numbers being given as "abstract" objects, our remarks in Chapter 2 should be recalled. It was pointed out there that Husserl does not wish to either psychologically or metaphysically "hypostatize" abstract objects. We cannot construe the talk of structure here in a naive conceptualist way, nor can we view it as naive platonism. One immediate reason for this is that our remarks about form are made in the context of the phenomenological reduction. I shall not attempt to say any more about this here however.

In Chapter 4 we noted that the lack of sensory data for mathematical

objects gives rise to what was called the plasticity objection to mathematical intuition. It should be clear how we can escape this objection. Acts of number intuition are nonplastic since they must be determined by the particular given underlying process of collecting that is carried out, i.e., by the *given* multitude that we run through. The relevant "given" in ordinary perception is comprised of sensory data. But the relevant "given" in acts of number intuition consists of the particular underlying process that is carried out, and what is relevant about this process for determination or knowledge of number is its structural features, not the qualitative features associated with the acts involved. If we try to analyze an ordinary example of making a number determination into its distinct components we would find that collecting, reflecting, and abstracting would all be involved. Husserl's claim would be that reflecting on collecting would not by itself yield intuition of numbers for one would not then necessarily be aware of the same structure across different cases, or of the inessential character of qualitative aspects of the objects numbered to the task of making the determination of number. Hence, we might say that acts of reflection which have the particular underlying processes of collecting as their objects will provide the material that gets "interpreted" in acts of formal abstraction, yielding intuition of abstract objects. We can evidently "collect" objects without reflecting on what we are doing, and we can also evidently reflect on what we are doing without abstracting, but there would be no intuition of numbers were there not acts of collecting, reflecting and abstracting (also comparing).[23]

Hence there is a kind of "given" in acts of mathematical intuition that we could regard as the analog of representing content insofar as it constrains and determines these acts. The process of constructing four units, for example, is different from the process of constructing five units. The structure of the former process is different from the structure of the latter process. The founded acts will be constrained and determined by just what is given in the founding acts. We determine the structure of just what is given in the founding acts and we cannot alter what is given about the founding acts at will. The founding acts thus determine which object will be individuated in the founded acts. An even more striking example of this will be discussed in the next chapter. It seems to me that the kind of view we have put forward about the construction of new objects being founded on what is previously given is essential not only to the details of the phenomenological theory discussed here but to many versions of constructivism in mathematics, including intuitionism.

8. SYNTAX AND REPRESENTATIONS OF NUMBERS

The idea of intuiting or constructing a unit is thus to be understood in a very general way but in constructing a unit the determinable x's in the construction can be made quite determinate by choosing certain objects to be the x's. I have been using stroke inscriptions, which are objects of straightforward perception, to mark construction of units. In choosing objects to be the x's I shall say that we have an *instance* or a *representation* of the construction by way of the underlying acts. Instances or representations of constructions are given in applications of the concept of number. Instances can be made concrete by choosing concrete objects to be the x's. If we take the concrete objects to be sign tokens then we could use strings of strokes or any other signs as representatives of units. It might also be mentioned that the construction or intuition of a unit can be made determinate by imagining the unit as some particular object. It will not of course matter to the construction just how we imagine the objects.

It will be worthwhile to consider for a moment the case where we represent constructions by finite, space-time configurations of symbols. In particular, if we take strings of strokes to be the numerals of a formal system of number theory they will be representatives of the numbers they denote in the sense that |, | |, | | |, | | | | ... have the property that the nth is a configuration of n objects, and that it will be possible to reconstruct from it the sequence of numerals which precede it. Each numeral contains within itself all the previous ones so that the nth numeral is itself a representation of the numbers from 1 to n. Similar remarks about representation can be made in the case of standard numerals of formal systems of number theory where we begin with 0 and s or ′, for in this case the number represented by the numeral is the number of occurrences of the symbols minus one. The set of numerals generated by some procedure will be isomorphic to the natural numbers in the sense that it has an initial element and a successor relation. For this reason, following the view of logicians like Martin-Löf and Prawitz, we might regard formal numerals as *canonical* representations of the fundamental constructions of which we have been speaking. In their case, unlike the case of Arabic, Roman, or other systems of representation, no computational decomposition is required in order to determine which number a numeral represents. Contrast this with the case of an Arabic numeral like 1986 which would naturally be understood in terms of addition, multiplication and exponentiation as $103 + (102 \times 9) + (10 + 8) + 6$. Of course for large numbers the formal numerals increase in length too rapidly to be practical. Since

Arabic numerals increase in length more slowly they can represent much larger numbers by easily surveyable numerals. But even Arabic numerals are not perspicuous when we consider numerals of, say, twenty digits. Hence we devise other expressions in this case for more perspicuity. We shall have more to say below about the computational decomposition or composition involved in number constructions.

While it is possible to represent constructions by finite, concrete configurations of signs in various ways the construction itself is of course not to be viewed as a syntactical object. Rather, the construction is an experience or mental process that we carry out. The syntactical representation has quite different properties from the construction. If we look at the representation in terms of sign tokens then it is given in a certain kind of ink, has a particular color, and so on. These properties are obviously irrelevant to number intentions. A similar point can be made if we look at the representation as a sign type because the properties of sign types will not be relevant to number intentions. The particular properties that a representation has will be by-products of our effort to make the intuition or construction of a unit concrete and determinate.

In viewing the matter of the intuition of numbers this way I do not wish to deny that the use of symbols has a very important role in our arithmetical knowledge. For symbolic representations of constructions are no doubt essential for keeping track of and communicating constructions, and they have a special role in computation and in the case of large numbers. A system of notation of some kind is no doubt a condition necessary for awareness of large numbers and for general reasoning about numbers. However, what is crucial phenomenologically is that a construction be viewed as a mental (intentional) process. For apart from this process numerals are merely meaningless marks on paper. In order for them to have any meaning we must approach them with certain intentions. As mentioned earlier, in the *LI* Husserl argued quite generally that a string of symbols counts as meaningful if and only if it is accompanied by a mental intention.

It should be clear from these last remarks that our account is different in some important respects from a Hilbertian view of the objects of intuition. We do not want to say that sign configurations are the *objects* of intuition in intentions directed to particular numbers. They may play a role in intuition as representatives of numbers in underlying acts but *numbers* are the objects of intuition. Numbers have quite different properties from numerals. Knowledge

about the former cannot be identified with knowledge about the latter. Numbers could not of course be objects of the same type as perceptual objects. They are not located in space-time, do not have color, shape, and so on.

CHAPTER 6

NATURAL NUMBERS II

1. INTRODUCTION

In this chapter we shall discuss several matters about the construction process for natural numbers that have not yet been mentioned: first, alleged difficulties about the numbers 0 and 1; second, questions about constructions involving arithmetical functions, in particular '+' and '.', and the relation '='. We shall then briefly summarize our account of the processes involved in fulfillments of intentions to small numbers and singular statements about them and move on to questions about large numbers and general statements about numbers.

The same processes involved in intuition of quite small natural numbers are involved in intuition of large natural numbers except that in the latter case we cannot actually obtain the object as a result of the process of construction. Our discussion thus far has suggested a modal principle about intuition of parts and wholes of objects: If we can actually complete the intuition of an object then it is possible to intuit the object. The converse would not hold. For large enough numbers the process cannot be completed. But, given the intention, we can at least actually obtain a part. It will be argued that our knowledge about large numbers has an intuitive foundation because we can actually intuit a part. This suggests a second, less trivial modal principle: If we can actually intuit a part of an object then it is possible to intuit the object. The converse of this principle should hold. It seems to me that this idea offers a new perspective on issues concerning mathematical intuition in the philosophy of mathematics and that its consequences could prove to be fruitful for developing a better understanding of mathematical knowledge. The intuition of the object in this case will be founded on intuition of its "parts", just as intuition of a medium-sized physical object is founded on intuition of only some of its parts. The processes which produce M's belief in the existence of large numbers are the same as those which produce M's belief in the existence of quite small numbers, but the evidence given in the former case must surely count as different in certain respects from the evidence given in the latter case. The difference will be briefly discussed.

Issues concerning intuition of large numbers are closely related to issues

about the role of intuition in the knowledge that we have of statements proved by mathematical induction. In the case of mathematical induction we are not of course concerned at all with the question of the particular object that is obtained as a result of the construction process but only with the construction process itself. In induction no particular object (number) is intended. The relevant question is whether the intention expressed by a statement like $(\forall x)(\forall y)(x + y = y + x)$ is fulfillable for arbitrary m and n. The intuitive content of *general* statements about natural numbers will be founded on insights into the relevant construction processes for natural numbers. The insights involved will be much the same as insights that particular rules or humanly computable functions can be carried out and that they behave in a certain way.

In the latter part of the chapter we shall consider some processes which produce M's beliefs that the intentions expressed by such statements are fulfillable. Hence, we shall be considering the evidence for beliefs that general statements about numbers are fulfillable. The beliefs in these cases are beliefs about the construction processes, and so we are considering construction processes as objects. Acts of reflection will therefore be involved in these beliefs. It will be argued that M's beliefs that such intentions are fulfillable are founded on actually carrying out finite initial segments of constructions in which certain properties of the construction process for natural numbers can be recognized. In these cases acts of reflection which have as their intentions that "it is possible to continue the construction" can be fulfilled (or at least partially fulfilled) by actually carrying out the construction. In actually carrying out the construction it is seen that what was thought to be only "possible" at some early stage in the construction process becomes actual at some later stage. That is, what at some stage is a future or expected intuition becomes an actual intuition, and then a past intuition. Hence what was thought to be possible at the earlier stage really was possible. It was not just conjecture or hypothesis without the possibility of fulfillment. This provides the basis for a view about what is "possible" in intuition. It will be suggested that the question of where to draw the line on what is "possible" in intuition, apart from some general conditions that clearly set limits, can be viewed as a function of the degree of evidence one is prepared to accept. Similar observations hold in the case of sequences of intuitions of ordinary perceptual objects. Actually carrying out some sequence provides a foundation for the belief that it is possible to continue with an arbitrary number of acts, but there are different degrees and types of evidence involved.

These and other observations to be made later in the chapter have a bearing on some of the objections that have been raised to the view that the knowledge we have by mathematical induction has an intuitive foundation. We shall briefly take this up at the end of the chapter.

2. 0 AND 1

Problems about the numbers 0 and 1 have traditionally been raised for the kind of account we are giving. The construction of numbers is supposed to amount to the construction of determinate multitudes. The awareness of number is said to arise through specification of multitude. But the numbers 0 and 1 could not count as specific multitudes. Husserl wants to say that the numbers 0 and 1 are negative answers to the question "how many x's are there such that Sx?". In the case of 0 this seems quite plausible, but it is somewhat odd in the case of the number 1. The construction of the number 1 should pose no problem for our account of the intuition of numbers, for the fulfillment of the intention that "there is one object such that Sx" will just amount to construction of a unit. The number 1 is just the limiting case of the notion of a "multitude". We might look at the matter in the following way: The fulfillment of the intention that "there is one x such that Sx" just amounts to the fulfillment of the intention that "there is at least one x such that Sx" and the frustration of the intention that "there is more than one x such that Sx",

$$(\exists x_1)(\exists x_2)(Sx_1 \wedge Sx_2 \wedge x_1 \neq x_2) \rightarrow \bot$$

which is just

$$\neg(\exists x_1)(\exists x_2)(Sx_1 \wedge Sx_2 \wedge x_1 \neq x_2).$$

Perhaps what Husserl meant by saying that the number 1 arises as a negative answer to the question "how many?" is captured by looking at the matter this way. For if we approach a situation with the intention or belief that "there is more than one x such that Sx" i.e., that there is a "multitude" or a "many", this intention is frustrated in the case where there is only one object.

What about the number 0? A somewhat similar observation can be made. If we approach a situation with the intention that $(\exists x)Sx$ i.e., that "some" or "a multitude" of x's are S, and the intention is frustrated so that we have $(\exists x)Sx \rightarrow \bot$, then the intention expressed by $\neg(\exists x)Sx$ is fulfilled. One finds, for example, that one's belief that $(\exists x)Sx$ is absurd. In this case the intention that there are no objects such that Sx is fulfilled, so that there is a foundation

in our experience for what we call the number 0, but there is not intuition of 0. One does not collect anything here. Proof that $\neg(\exists x)Sx$ will involve the construction process for natural numbers. We might thus view 0 as an addition to the system of natural numbers as it is originally given. Historically, 0 did not enter the scene until quite late, and then only with some very odd superstitions attached to it.

3. NUMBERS FORMED BY ARITHMETIC OPERATIONS

We should also say something now about the intuition of numbers formed by various arithmetic operations, in particular, addition, multiplication, and exponentiation. How are we to understand intuition of numbers like $5 + 2$, $5 \cdot 2$, 5^2, and so on? In the case of '+' we simply construct five units, construct two units, combine the constructions and determine the number of units in the new construction. What it means for the intention expressed by '5 + 2' to be fulfilled could also be understood in the following way. We carry out a construction which gives the number 5, then we continue the construction by iterating the process of constructing a unit twice, i.e., we add two more units. Thus, addition of two numbers n and m can be understood as built up from the basic operation of adding one more unit, where this process is simply iterated. A number like $5 \cdot 2$, or more generally, $n \cdot m$, is constructed by iteration of addition of two numbers: $5 \cdot 2$ is just $5 + 5$. A number like 5^2 is constructed by simply iterating multiplication: $5 \cdot 5$. In *LI* Husserl actually suggests this kind of view directly.[1] He says that a number like 5^2 arises when one forms the product $5 \cdot 5$, where we can think of 5 as built up from the constructions $5 = 4 + 1$, $4 = 3 + 1$, $3 = 2 + 1$, $2 = 1 + 1$, where successive substitutions are made:

$$3 = 1 + 1 + 1$$
$$4 = 1 + 1 + 1 + 1$$
$$5 = 1 + 1 + 1 + 1 + 1$$

The construction thus consists at the bottom of a string of 1's and +'s. What this amounts to of course is simply viewing arithmetic operations as built up on the basis of the process of iterating a successor function, that is, on the basis of the fundamental kind of construction we have been discussing. This is how we would think of their definitions via recursion equations. Husserl says that in carrying out this construction

> It is plain that an act of fulfillment would correspond not only to this final result, but to each individual step leading from one expression of this number, to the expression next in order, which clarified and enriched its content. In this manner each ordinary decimal number points to a possible chain of fulfillments, whose links are one less in number than the number of their component units, so that chains of indefinitely many numbers are possible *a priori*.

To intuit 5^2 as a whole would be to complete this process. If the process is not completed one intuits only part of the number. But in any event, Husserl says that it is a property of cases of this sort that "in them the *content* of the presentations – or, more clearly, their 'matter' – *dictates a determinate order of fulfillment a priori.*" I suggest that this claim be read as saying that the order of fulfillment is determined *a priori* because the sequence is *lawlike* or determined by a rule. For by virtue of lawlikeness future intuitions are determined in advance of our actual experience. This surely counts as a priori knowledge in at least one sense of the term. As I said earlier, I think this is a distinctively Husserlian conception of the a priori.

As we mentioned, recursion equations could be used to characterize the various constructions that could be obtained on the basis of the construction process described in the last chapter. In using the recursion equations we would need to see intuitively that they hold. This would be analogous to seeing intuitively that mathematical induction holds and we shall take this up in a moment. Recursion equations contain '=' and so the process of comparing would be involved. In fact, comparing will generally be involved in singular statements of the following forms: $2 + 3 = 5$, $2 + 3 > 4$, and so on. Note that to intuit *that* $2 + 3 = 5$ involves construction of $2 + 3$, 5, and a comparison.

In concentrating on recursion equations for addition, multiplication, and exponentiation I do not mean to suggest that historically our understanding of these operations sprang into existence in just this form. Rather, since I am trying to do justice to arithmetic as we now know it, to the intentions or cognitive acts of someone who thinks about arithmetic at the present time, I am suggesting that we view recursion equations as clearly codifying processes for fulfilling intentions to the numbers referred to by terms in which these operations occur. This is no doubt a later historical development in the understanding of these operations, but then I would argue that one could view the historical development of arithmetical knowledge generally in terms of the genetic development of cognition and corresponding intuitions.

4. SMALL NUMBERS AND SINGULAR STATEMENTS ABOUT THEM

The remarks made up to this point should suffice to indicate how we can understand intuition *of* particular small natural numbers and intuition *that* in the case of singular statements about them. To intuit a natural number n, as we have described it so far, is to carry out a construction up to n. The construction is a process which gives evidence for the belief that the number exists. This is how the idea of fulfillments of intentions to objects is generally to be understood. Intuition consists of being able to have a sequence of mental acts in which the object is given. To do something so elementary as to verify to consciousness the existence of the number 3, for example, is to carry out this process up to the point at which the number 3 is given. From what has been said this amounts to constructing a unit, and then iterating the process of constructing a unit until we arrive at the number intended. The process in which the number is given can be repeated in different acts. It will be involved in the intuition of each particular number n, and it may be viewed as underlying M's belief that S, where S is a singular statement about small numbers containing arithmetical predicates.

Now on what grounds can we say that we have given an account of intuition *of* numbers? We can say that the *processes* we have described are processes that are involved in fulfillment of intentions expressed by the singular terms of arithmetic, just as certain processes would be involved in the fulfillment of intentions expressed by singular terms from discourse about ordinary perceptual objects. In both cases we should be able to speak, at least phenomenologically, of objects. What this means is that we are speaking about objects obtained as a result of a construction process. The objects of perceptual intuition are of course of a different type from objects of number-theoretic intuition. We can also readily contrast intuition *of* numbers with intuiting *that* some state of affairs obtains for some particular numbers. The processes involved in each case would be different. Intuiting that $2 + 3 = 5$ would involve construction of 2, 3, $2 + 3$, 5 and a comparison. The comparison would be explicitly called for to verify the identity.

It should be noted that in our description of the intuition of natural numbers we have not tried to identify natural numbers with any particular objects, e.g., von Neumann sets, Zermelo sets, strings of strokes. Rather, numbers can be represented in any of a number of ways. It would be a mistake to identify numbers with certain objects at the expense of any other objects for which arithmetical operations would be defined and laws of arithmetic preserved. Our remarks in Chapter 4 about structuralism can

already be filled out to some extent. We can agree with the structuralist view that there is nothing about natural numbers that is not determined simply by their structural relations to one another. Our work suggests that it would be useful to distinguish ordinal from cardinal structuralism. The number that will be reached at any stage in the construction process will be indeterminate or "incomplete" with respect to a reductive definition in the sense that one cannot suppose it will be some particular object and not some other particular object, a von Neumann set, say, and not a Zermelo set. We cannot think of each number term as having some unique referent in this way: the number cannot be identified with a particular representation of it. This should be clear from our discussion of the indexicality of intuition for it is this feature of the intuitive process that permits different construals of natural numbers.

If we recognize a difference between mathematical and perceptual objects we should not of course expect that mathematical intuition will give objects that are complete and individually identifiable in the same way that perceptual intuition gives objects that are complete and individually identifiable. From what we have said about the intuition of natural numbers it is clear that each natural number will be determined by its place in the sequence of natural numbers. We are able to fully determine places of numbers in a sequence by carrying out some construction up to n, and by the process of comparing. By carrying out constructions and comparing we know that there is an ordering among numbers and that places in the sequences are fixed. It is this process which permits us to individually identify numbers, and the identification can be made only by determining where numbers stand in relation to one another. There are no other criteria that could be used to individuate numbers that would not attribute extraneous properties to them. On our view of intuition, therefore, we do individually identify natural numbers in intuition. We can give an account of individual identification that is appropriate to these objects. It is in this way that we have the means for individually identifying the objects to which the intentions expressed by number terms refer.

While the objects of perceptual and number-theoretic intuition are different the processes involved in each case are analogous. For in either case individuation of an object will be a function of a sequence of indexical mental acts. It will be from the sequence and the rules governing the sequence that what is known at any given stage will derive its individuality. What is involved in this structure in perceptual intuition is sensory input and this input will make perceptual intuition concrete and definite in a way that mathematical intuition will typically not be. Sensory data play a role in the

individuation of perceptual objects in intuition that they do not play in the case of mathematical objects. For natural numbers however we can find something akin to the concreteness and definiteness insofar as it is possible to choose concrete, perceptible objects, like sign tokens, to represent numbers.

While we cannot suppose that we intuit natural numbers as "complete" in the same way that we intuit perceptual objects as complete, there is a sense appropriate to mathematical intuition in which we can say that the natural numbers given by our constructions are complete. This has long been recognized in intuitionistic mathematics. We cannot suppose that the referents of number terms can be fixed in any more determinate way by associating them with certain objects at the expense of any other objects that would do. But once we carry out a construction up to some n, as this has been described, we can regard the construction of n as complete. Nothing could be added to it without endowing numbers with properties that are alien to them as numbers. This kind of completeness is not trivial since it can be contrasted, for example, with the construction of real numbers where we suppose that real numbers are generated by "free choice sequences". Free choice sequences may be thought of as sequences which are such that at any given stage some total or partial restriction on future choices of values may be specified by laying down some law.[2] What is determined about the sequence at a given stage in the process of generation is what is known from the initial segment of it up to that stage, and the laws that have been specified. Then if we think of real numbers as generated by such sequences, there will clearly be no stage at which the object – some real number – is completely described. At each stage in the process of generating values the object itself is being generated, and at no stage in the process is the generation complete. In this sense, the intuition of natural numbers can be described as an intuition of individuals, and at any given stage it gives a complete object.

What has so far been said about natural numbers as objects of intuition should be compared with our earlier remarks on physical objects and the phenomenological reduction or epoché. From the point of view of phenomenology we do not need to consider anything more than the structure of the acts and sequences of acts in which objects are given, whether the objects in question be mathematical or physical. We are not to suppose that there is some kind of relation independent of this process which puts us in touch with objects. What phenomenology attempts to do is to analyze the act-structure which is necessary for knowledge of objects. We have described this structure in some detail in the case of the natural numbers. A question that we could now raise is whether the account we have so far given of the

intuition of natural numbers is compatible with the view that natural numbers are mind-independent objects or rather, whether they are to be viewed as our own mental constructions so that they could not be taken to exist independently of these constructions. We have been saying throughout this essay that intentions may be thought of as referring not only to objects that might exist independently of us but also to construction processes in which objects would be given. There seems to be no reason why we could not adopt an attitude which accommodates both of these views. Husserl's remarks strongly suggest that the idea that natural numbers are mind-independent objects is compatible with a phenomenological analysis of the intuition of such objects. The reason for this is that in phenomenology we are attempting to analyze sequences of acts by which we obtain knowledge of objects, independently of naive metaphysical questions about the objects. To investigate constructions does not require that we take a stand one way or the other on the question of the existence of numbers independently of the way they are given to consciousness. As we depicted the phenomenological reduction it would not be claimed that the existence of a number depends on the possibility of giving a construction, but rather that the intuition or consciousness of the objects as existing does indeed depend on having a construction. Being able to give a construction explains our ability to have knowledge of numbers.

This seems to be a position which is close to intuitionism in certain respects, especially as it is presented in Heyting's writings. Both Brouwer and Heyting say that the intuitionist program consists of the investigation of mental mathematical constructions as such, without reference to questions regarding the nature of the constructed objects, such as whether they exist independently of our knowledge of them. Heyting says that such a question is a metaphysical question and it is not the business of mathematics to investigate metaphysical questions. When pressed further on the matter in his *Intuitionism*, Heyting denies that intuitionism is committed to the view that mathematical objects come into existence only with construction, but he refrains from further comment. This kind of methodological restraint seems to be much like Husserl's position, especially in his later writings. The parity of this kind of position with respect to both mathematical and physical reality is also reminiscent of some of Gödel's remarks. I shall comment on the matter further in Chapter 8 where issues about constructivism and platonism in mathematics are discussed once again.

Let us make a final remark about the nature of the evidence the construction process we have described can be presumed to give in the case of quite small natural numbers. We shall then consider the construction process in the

case of large numbers and in mathematical induction. Obviously in the case of quite small natural numbers the construction process can actually be completed. The evidence we have for the existence of such numbers would presumably count as adequate, for all their "parts" are presented in intuition and the intuition is not bound to be underdetermined as in the case of quite large numbers. The evidence should perhaps also count as apodictic since we cannot imagine what it would be like not to be able to carry out the construction in these cases. Certainly the circumstances under which we could not carry out such constructions, under which it would be impossible, would have to be quite pathological. Note that in saying this we do not mean that we know every possible fact about such numbers, or every possible relationship they may be involved in. We are only speaking about fulfillment of intentions expressed by singular number terms, not about fulfillment of intentions concerning properties or relations of such numbers. The process we have described is also repeatable at different times and places and by different people. Repetition of a given process produces the same object. The intersubjective dimension of evidence that is entailed by the Kantian strategy is reinforced by the possibility of communicating constructions for small natural numbers by way of syntactical representations that are completely surveyable in straightforward perception. I shall have more to say about the lawlikeness of the process in a moment.

5. LARGE NUMBERS AND MATHEMATICAL INDUCTION

We have so far concentrated on the construction process for natural numbers in the case of quite small numbers. For small numbers the construction process can actually be completed. We shall now make some remarks about the construction process in the case of large numbers and in proofs by mathematical induction. In doing so we can give some important indications about how our treatment would lead to a satisfactory account of arithmetical evidence in general. We shall treat large numbers and induction in tandem since what is important in both cases is the insight that it is *possible* to continue the construction process and the knowledge of how the process would go. We shall not try to prove that the induction principle holds on the basis of the finite number of individual numerical intuitions we can actually have. Any such attempt would be circular. Rather, we shall concentrate on how the belief that it is possible to continue the construction process is built up, and the most important observation we want to make in this connection is that conclusions by induction "transcend" our actual experience, just as

maximization of perceptual knowledge transcends our actual experience. Our knowledge of conclusions obtained by induction is thus not "absolute", but rather has the character of being a belief for which we possess a certain type or degree of evidence. The main point we want to make then concerns the analogy we have been drawing: just as we suppose that perceptual intuition yields knowledge even though we cannot actually maximize our knowledge so we should also be able to say that number-theoretic intuition yields knowledge even though we cannot actually have all the intuitions necessary to maximize such knowledge.

In the case of large natural numbers we of course have intentions directed to some particular number, while this is not the case in proofs by mathematical induction. In the latter case there is no particular intended output of the construction process. Rather, in the case of proofs of *general* statements, like $(\forall x)(\forall y)(x + y = y + x)$, it must be seen that the intention expressed by the statement is fulfillable for arbitrary m and n. I would argue that we can generally regard such intentions as fulfillable when we possess a construction process (rule) for obtaining the objects to which the predicates apply. For the purposes at hand we might think of such construction processes or rules as algorithms or (humanly) computable functions. The predicates in the induction principle could be quite elementary, eg., primitive recursive, but we should generally be prepared to recognize an intuitive component in induction so long as the predicates involved are "constructive". This seems a reasonable assumption to make even if we do not have in advance a very clear conception of the extent of the universe of constructions. This does however evidently rule out as founded on intuition a classical second-order version of induction. If we looked at the matter in terms of the Kolmogorov interpretation we would want to know that the problems expressed by general statements were solvable given an arbitrary input. We could regard such problems as solvable when we possessed a (person) program of the relevant type. We can thus understand general statements much as in intuitionistic developments, with the qualification that we do not want to be uncritical about the evidence we have for these statements.

Since we have appealed to the notion of a rule (or a "function given as a rule") we should make a few brief comments about rules. We could think of rules as programs, that is, as finite, explicit (for a human) lists of instructions. The notion of a rule in this sense should be taken as primitive, not as a defined or definable notion. The notion of a rule is thus to be understood, as in the case of other primitives, in terms of an analysis of its origins. A function that can be given by a rule in this sense admits of a complete

description, unlike, for example, choice sequences. We could then identify and distinguish such rules or programs, and determine other properties and relations of programs. Since different programs may compute the same function we would want to recognize a distinction between intensional and extensional identity for programs or rules. There are many details about rules that could be gone into but for the purpose at hand it should suffice to say that the intentions expressed by general statements about natural numbers and finite sets could not be regarded as fulfillable if we did not possess the kind of construction processes provided by such rules. For certain parts of mathematics we obviously do not possess such construction processes. So it is the existence of the process that is essential. It is of course an assumption of this view that we can recognize such a process or rule when we see one, and that we can determine that it works. I do not wish to exclude the possibility that we could be mistaken about such matters, for it might happen in the development of mathematics that acts in which we believe that we do (or do not) possess such a process come to be corrected through further acts, i.e., through further intuition. So at a given time we have better or worse evidence for the existence of such a process and for the fact that it works.

In saying this we are taking note of the distinction drawn by Parsons between the kind of evidence we have for mathematical objects like natural numbers and the kind of evidence we have for mathematical induction and other principles of the same kind. For the insight into the possible progression of the generation of intuitions according to a rule *is* different from the intuition of particular natural numbers. In the case of intuition of natural numbers it is the numbers that are the objects of intuition, but the evidence we have for mathematical induction concerns general propositions about natural numbers. In the latter case it is construction processes for natural numbers that are the objects of our acts (of reflection). We can readily admit that the mathematical knowledge involved in the latter case has a "complex relation to 'intuition'" as understood in the former case. In the latter case we are concerned with fulfillment of acts of reflection directed to construction processes, while in the former case we are concerned with fulfillment of acts directed to particular numbers. In the intuition of particular numbers we simply carry out the construction process until we reach the number intended. Note that on our view the knowledge that we have by such acts of reflection is to be regarded as having an intuitive component. As we noted in Chapter 1 it is unclear whether Brouwer means to have such a view, and likewise, whether Kreisel and Feferman do in their theories. It seems to me that intuitionistic mathematics would probably be philosophically committed to

the view that reflection does involve a kind of intuition.

6. THE POSSIBILITIES OF INTUITION

Perhaps the most important element in the discussion of intuition and induction, in light of the objections mentioned in Chapter 4, concerns the belief that it is "possible" to carry out the relevant construction processes. We should therefore make some further observations about this belief. Let us recall our remarks on knowledge in Chapter 1 and begin by concentrating on the belief that it is possible to continue the basic construction process that we described in the last chapter, and on certain other beliefs about how this construction process behaves. In accordance with our remarks on knowledge we argue that these beliefs are produced by a process which gives evidence for them and that, just as anywhere else, it is *intuition* that gives evidence for our beliefs. Since the beliefs here are beliefs *about* the construction process we are obviously speaking about intuition in the case of certain acts of reflection. Our beliefs about the construction process could either be fulfillable with respect to the process or not. It is only though intuition that we can decide which beliefs about the construction process are correct.

The place to begin in considering the origin of the belief that it is possible to continue the basic construction process is evidently with quite small finite constructions because the belief would no doubt be founded on actually carrying out such constructions in which certain properties of the construction process can be recognized. Let us consider again the following representation of a construction of the kind we have been discussing (see Fig. 2),

| | | | | | | | |

As we run through the strokes here there is one act directed to each stroke. If we think of Husserl's notion of the external horizon of an act then at a given stage in the process the stroke at that stage is in the foreground while the other strokes are in the background. At the next stage in the process a stroke that was in the background at the previous stage is brought to the foreground. At an early stage in carrying out the construction, let us say at t_3, I have taken into account a number of strokes but the remaining strokes will be protended. I anticipate running through them at future stages. The awareness that there are further strokes to be taken into account is part of my act at t_3. They are, in Husserl's terminology, in the horizon of my act at t_3.

We noted in Chapter 3 how in *EJ* Husserl spoke of the horizon associated with an act as the "induction which belongs essentially to every experience

and is inseparable from it, being in the experience itself." Husserl says in *EJ* that the term "induction" is useful because is suggests "induction in the ordinary sense of a mode of inference and also because it implies that the latter, for its elucidation to be completely intelligible, must refer back to the original, basic anticipation." Mathematical induction also concerns just such anticipation. For in an act at a given stage we anticipate that we *can* construct another unit. This kind of anticipation will be part of any act in the sequence. In fact, it seems evident that while we can actually carry out only finite sequences of intuitions we would always know, by virtue of the act's horizon at any given stage, how to complete one more step in the process. The act's horizon, it should be recalled, is determined by the intention of the act and by relevant background beliefs. Conclusions of proofs by induction on this view are therefore statements about what can be expected, about future intuitions. This is much like Bishop's view in recent writings that general mathematical statements are like predictions based on a finite amount of data.[3]

The belief that it is possible to continue the construction is evidently built up on the basis of small finite constructions of this type. For at t_3, for example, it is clear that I *can* take account of the next stroke. It is in this very elementary kind of anticipation that the modal notion has its origin. I protend the other strokes and it is then a matter of actually taking them into account. The knowledge that I can construct further units at a given stage will be based on what I have been able to do in acts up to that stage. At t_4, for example, my knowledge that I can construct another unit is founded on seeing that at t_4 I have actually constructed a unit that at t_3 was only possible or only anticipated. And at t_3 I actually constructed a unit that was at t_2 only possible or anticipated, and so on. The knowledge of what is possible with respect to carrying out the construction is founded on what I have been able to do in the past. In the very process of carrying out the construction we experience that it is possible to continue carrying out the construction. Having completed some construction process we see that we actually constructed a certain number of units that were at t_1 units that it was only possible to construct.

The belief that it is possible to continue the construction process can thus be fulfilled in small finite cases by actually carrying out the construction. This amounts to saying that acts of reflection on the construction process which have as their intention that "it is possible to continue the construction process" can be fulfilled. In actually carrying out the construction it is seen that what was thought to be only possible at some early stage in the construction becomes actual at some later stage. Thus, what was merely thought or

intended to be possible at the earlier stage really was possible – it was not mere conjecture without fulfillment. The belief that it is possible to continue the process is thus fulfillable with respect to the process. If it were not we would not be entitled to regard it as knowledge.

It is useful to ask under just what conditions it would not be possible to take account of the next stroke or strokes. Clearly in the case of small finite constructions the conditions would have to be quite pathological, just as they would in ordinary sense-perception if we could not undertake further perceptions of an object or objects. In the case of small finite constructions we are convinced that it is possible to continue the construction process because even the contingencies involved in preventing us from actually carrying out very large constructions – life is too short, etc. – are not involved here. In the case of large constructions these contingencies are involved, so it might be argued that it is not possible to carry out these constructions, and hence that it is not possible to intuit large numbers. The notion of possibility on which this argument is pegged is one that depends on our ability to actually complete a construction. It is a notion of "practical possibility", or "feasibility". What I would like to say about it is that it is a notion of possibility that results from making one of a variety of determinations that could be made about evidence. But from what we have been saying we do not have this kind of evidence in ordinary perceptual intuition because we intuit only parts of objects, and we cannot actually complete the intuition. Even so, we do have knowledge about such objects. So if we relax the conditions on evidence somewhat we obtain a broader notion of possibility, one that does not depend on actual completion but rather depends on being able to actually intuit a part of an object and having some insight into how to continue the intuition.

This brings us some distance toward the kind of "in principle" possibility that is often discussed in constructive mathematics. This notion of possibility depends on being able to have partial intuitions, but it also depends on some observations about the *structure* of consciousness of just the kind we have been making. Given what we have said about the retentive-protentive structure of the immediate horizon of consciousness what it would mean for it not to be possible to continue the construction process is that the retentive-protentive structure of consciousness would have to be different from what it is, or that it could not exist in the form that it does. If, for example, the protentive awareness of a further unit or units were not part of my act at t3 we would perhaps not know how to continue the construction, or we would not know how to continue it in just the way that we do. If we reflect on the

structure of consciousness here the point is general. The particular numerical subscripts can be dropped. At an *arbitrary* stage t_n we would not know how to construct the object at t_{n+1}.

It is sometimes said that the belief that it is "possible" to continue a construction in this sense involves a notion of possibility according to which various biological, historical, psychological and other contingencies of particular individuals are irrelevant. We are not concerned, that is, with what is possible for particular individuals located at particular times and places, in particular situations, having particular personalities, and so on. As we said earlier, the mathematician M in our analysis must ultimately be understood not as the contingent biological or psychological being but rather as what Husserl would call the "transcendental ego", or in a different tradition, what intuitionists might call the "idealized mathematician". To ask what is possible for the idealized mathematician is to ask about the structure of consciousness of the type of beings that we are. It is to consider the type of consciousness of which we are individual instances. To those who are uncomfortable with the notion of a transcendental ego it might be helpful to think of the idealization here as similar to the idealization involved in arriving at the notion of a Turing machine. (Of course we do not necessarily mean to identify transcendental egos with Turing machines.)

Another way of thinking of talk of the idealized mathematician or transcendental ego is in terms of abstraction from what can *actually* be completed to what can *possibly* be completed in the case of intuition of large numbers and induction. We are really concerned with the *form* of the construction process. Note that abstraction from what can actually be completed is quite different from the kind of abstraction involved in supposing that the construction of numbers is actually completed. The latter kind of abstraction is involved in the conception of the actual infinite.

As we mentioned earlier, the type of consciousness of which we are individual instances might be distinguishable from other possible types. Actually completing an infinite sequence of intuitions is not possible for us (for beings of M's type). We might try to be more definite about what is possible with respect to M's constructions by drawing some comparisons based on relaxing conditions in various ways on what is to count as intuitable. One way of relaxing these conditions would amount to allowing that once a process has been specified in such a way that each individual step could be performed by M it could be *postulated* that the process is completed, even though M could not himself complete it. In this case we are in effect separating the output of the process from the process in which it is generated

and then treating it as a single completed object. Then the generation of the set of natural numbers, for example, could be regarded as actually completed. The being who could complete this process would be able to actually synthesize an infinite manifold (or multitude). We could postulate even more powerful intuitive abilities by relaxing the condition that the processes involved be carried out in discrete steps in time. For then we might conceive of a being who could run through any number of steps in a finite time, thus actually synthesizing larger and larger infinite manifolds. We have then of course abstracted in a fairly substantial way from M's intuitive abilities. I shall have more to say about these in matters in Chapter 7.

Reflecting on the basic construction process it can be readily recognized that the process is carried out according to a very simple rule: iteration of a successor operation. As we have described the construction of the natural numbers no properties of numbers are used beyond identity and non-identity, and the generation of numbers will be well-determined in the sense that there is no choice about how to extend any given initial segment of the construction. The sequence in which we construct the numbers is completely fixed in advance by a rule so that there is no difficulty about knowing how the sequence would go if it were to be continued. Our belief that the process is lawlike is founded on these kinds of considerations about the construction process. We can actually work through different constructions and in each case we see the same thing. This gives a foundation for the belief that it is possible to continue the construction.

7. SUMMARY OF THE ARGUMENT FOR LARGE NUMBERS

Let us summarize our argument for speaking of intuition in the case of large natural numbers. In the case of large natural numbers M cannot of course actually obtain the number as a result of the process of construction. Because of obvious time and memory limitations on our actual constructions we can complete only some finite initial segment of the process of intuition in the case of large numbers. It does not follow, however, that M does not intuit large numbers. Intentions directed to large numbers can be at least partially fulfilled and on the basis of reflection on the construction process we have evidence for regarding them as fulfillable. The intuition of the number will be founded on intuition of its "parts" just as intuition of a medium-sized physical object is founded on intuition of its parts. In ordinary sense-perception we intuit an object toward which an intention is directed even though we are actually given only parts of the object. Background beliefs associated

with an act's intention in the case of either ordinary perceptual intuition or in arithmetical intuition, and the horizon determined by them, serve in various degrees to make up for indeterminacies in what is actually given in the intuition. The insight into the possibility of continuing the construction for natural numbers is analogous to the insight involved in seeing that we could continue ordinary perception of an object or objects. The actual incompleteability of the intuition in either case is one of the factors that leads to the phenomenological description of objects as "transcendent", and as appearing to be independent of consciousness, and it merely reflects the fact that we cannot have a maximal degree of evidence for the existence of the kinds of objects that require a very large or even an infinite number of partial perceptions for maximization of knowledge.

The degree of evidence we have for the existence of large numbers must obviously differ from that which we have in the case of quite small natural numbers. We can actually complete constructions for small numbers, but not for large numbers. The evidence would not be "adequate" and perhaps it would also not count as "apodictic". Other reasons for thinking that arithmetical knowledge in the case of very large numbers might not count as apodictic have to do with the existence of undecided arithmetical propositions like Fermat's Last Theorem, or Goldbach's conjecture, and are perhaps best discussed in connection with the kind of evidence we have for induction and related principles. The important thing to notice for our purposes, however, is that the evidence we have for the existence of large numbers is not disanalogous, as regards its adequacy, from the evidence that we have for medium-sized physical objects.

It might be pointed out that because we can speak of evidence for the existence of numbers as a matter of degree we need not feel committed to setting any definite bounds on what is to count as a "small" number. Evidence for smaller numbers shades off into evidence for larger numbers. While I shall not take up the matter here I think that our view will not be plagued by problems of the sort that Dummett raises for "strict finitism", problems about the coherence of notions like that of "feasibility".[4] We might simply view attempts to specify what is feasible as specifications of more or less stringent conditions on degrees of evidence concerning natural numbers. There may be a variety of motivations for choosing one set of conditions over another, but the epistemological question of just which set of conditions is the correct set if we are to have a secure and reliable foundation for mathematics would be out of place. No definite line could be drawn.

8. FURTHER COMMENTS ON MATHEMATICAL INDUCTION

Let us now make some further specific comments about the role of intuition in the knowledge of conclusions obtained by induction. On our view proofs by induction are to be thought of as simply paralleling the given construction process. We see that certain statements hold for natural numbers by carrying out the generating construction described earlier. The construction of a number n consists of successively building up numbers, the numbers from 1 to n. At a given stage in the construction we can stop to investigate whether the number obtained at that stage possesses a certain property or not. For example, we can determine whether a number m, where $m \neq n$, occurs in 1...n or not. In the first case we obviously say that $m < n$; in the second case that $m > n$. It is a simple theorem that if $m \neq n$ and $m > n$ then $n < m$. For if m does not occur in 1...n we know that at the stage where we reach n the construction of m is not complete. Thus, n occurs in 1...m. We can make a similar observation about proofs by mathematical induction. For suppose that S(x) is a predicate of natural numbers such that S(1) holds and that for every n, $S(n) \to S(n')$. Let m be any natural number. Running through 1...m we know that S, which holds for 1, will continue to hold at every stage in the construction of m. Therefore, S(m) holds.

We can claim that there is intuition that $(\forall x)Sx$ in a proof by induction because for the natural numbers we have a method of construction which, given any $n \in N$, would yield by specialization the construction Sn. On our view the intuition that $(\forall x)Sx$ does not depend on actually intuiting that Sn for each $n \in N$. The intuition that $(\forall x)Sx$ is founded on actually carrying out some of the intuitions along with the fact that we possess the relevant construction process. We might note that there seems to be an *a priori* character to the induction involved here that is not involved in empirical induction. For the future intuitions are evidently completely fixed in advance of our actual experience, whereas this would not be the case in ordinary sense-perception. This a priori character might lead one to suppose that the future intuitions are fixed with a kind of "necessity", whereas in empirical experience probabilistic elements are always involved. The induction in mathematics is nonetheless presumably *synthetic* in a Kantian sense. For while what will be given in the n + 1 intuition may be protended in the act at stage n it is not actually given. It is only implicitly or potentially given, so that it does not have the same kind of being as what is given at stage n. To actually carry it out would go beyond or add to what is given in the act at stage n. On the Kantian view the conclusion of the argument would not be

already contained in the premisses.[5]

What we are concerned with, it should be recalled, is not the knowledge that conclusions by induction are true in an absolute or "ideal" sense but rather with the knowledge that they are *fulfillable*. As we said earlier, from the point of view of a phenomenological elucidation of knowledge we may not think of mathematical truth in terms of idealized truth conditions but we are instead concerned with mathematical truth "within its horizons". One of the sources of the charge that circularity is involved in intuitive justifications of induction no doubt lies in the unchecked assumption that the knowledge that we have by intuition is or must be knowledge of idealized truth conditions. For it is thought that only in this way could we be said to really know that induction holds. The problem is then that we cannot know that much by intuition except on pain of circularity.

Let us now summarize our views on mathematical induction. To say that conclusions proved by mathematical induction are fulfillable means that there is an intuition process that a mathematician M could carry out. As we have been saying, there is a distinct notion of possibility here that should be viewed in terms of the horizon of expectations associated with an act at a given stage of a construction. The idea that we could "possibly" intuit a number m, where m may be any number, should be understood as analogous to the situation in ordinary perceptual intuition where we believe that it is "possible" to have an arbitrary finite number of perceptions of an object or objects even though we can actually intuit only parts of the object. The fact that we cannot actually complete the perception of an ordinary perceptual object does not signify that we do not intuit the object, but only that the intuition is partial. By the same token there can at best be partial fulfillment of the intentions expressed by general statements about natural numbers, but it does not follow that there is no intuitive foundation for such statements. On the basis of the analogy with perceptual intuition it does not follow that we do not know something by intuition because we have not had all the possible intuitions of it, or because we have not completed the intuition. On the contrary, we are entitled to suppose that we have intuition that statements of the form $(\forall x)Sx$ hold even though we are not actually able to intuit that Sn for each $n \in N$. We can say that we have intuitive knowledge even though we cannot have the infinitely many intuitions *of* or *that* that would be required for maximization of our knowledge. What is important about the case of very small numbers is that we see we can *actually* complete their construction. This is significant because if M cannot actually carry out some of the steps of a purported construction process we should not be prepared to regard the

process as one that could possibly be carried out by M. Perhaps we could regard it as a possibility for some other kind of being, but then we are not concerned with M's constructions. I think that on the basis of these kinds of observations we could say that the actual intuitions *of* are necessary for the knowledge that the intentions expressed by general statements are fulfillable. This is quite different however from saying that we only understand the general statement via the infinitely many instances of it, i.e., via infinitely many intuitions *of*.

On our view it is possible to avoid the kinds of objections about representing arbitrary numbers in intuition that were mentioned in the last chapter. For the problem with the concept of intuition discussed there was that it would be necessary to represent reasoning about arbitrary numbers by perceptible but "indeterminate" syntactical configurations, like '|...|'. If '|...|' is taken to have a particular number of strokes it is not arbitrary and hence we cannot see by intuition that induction holds for every numeral. But if '|...|' does not contain some particular number of strokes we could not see by intuition its internal structure that we would need to see to know that induction holds for it. On our view it is not necessary to suppose that we can represent arbitrary numbers by perceptible but indeterminate syntactical configurations. The intuitive foundation of induction depends instead on the fact that we possess a certain construction process, some steps of which we can actually complete. We understand the idea of an "arbitrary n" with respect to this process: an arbitrary n is an object obtained at some stage or other of the process.

The kind of evidence we have for mathematical induction and other principles of the same kind is thus different from the kind of evidence we have for the existence of particular numbers. We could only know that the intentions expressed by general statements about natural numbers are fulfillable by acts of reflection on the construction process. Acts of reflection on the construction process are themselves characterized by the difference between intention and fulfillment. They must be fulfillable if they are to give knowledge. This kind of intuition, as noted, is different from the intuition of particular natural numbers because it has the construction process as its object. From all that we have said about intentionality and intuition there would seem to be no reason to suppose that it is misleading to speak of processes involved in fulfillment of acts of reflection as intuitive.

We might again ask about the nature of the evidence we have for the belief that inductive conclusions are fulfillable. Is it adequate? Is it apodictic? It seems that it is neither. The main point I wish to note is that acts of reflection in general can no more be supposed to provide truth in an absolute sense than

any other acts. They can provide evidence but unless the evidence can be shown to be adequate, apodictic, and so on, they cannot be presumed infallible. The beliefs produced by acts of reflection might need to be corrected or more fully determined and this can only be done through further acts of reflection, through further experience, as we noted in Chapter 2. Further reflection would give us a better approximation to truth. The only way we could hope to avoid such further reflection would be to come to the realization that further reflection could not possibly make a difference to what is seen in a given act of reflection.

I would not be prepared to claim that arithmetical knowledge in general is either adequate or apodictic for several reasons. A primary reason is that there are obviously undecided propositions in elementary number theory. Such propositions might be either true or false. If arithmetical knowledge in general were apodictic or adequate this kind of incompleteness or indeterminacy in our knowledge could not exist. In any case, it seems we are certainly entitled to say that our knowledge that it is possible to continue the construction process for natural numbers is at least as reliable and secure as our knowledge that it is possible to carry out the process of intuition of medium-sized physical objects that would be required for maximization of knowledge.

9. INTUITION AND AXIOMS OF ELEMENTARY NUMBER THEORY

The foregoing considerations might be understood as a view about the intuitive foundations of our knowledge that the axioms of elementary number theory are true. This might be viewed as the analog for elementary number theory of Gödel's much stronger suggestion for set theory that it is on the basis of mathematical intuition that we see that the axioms of the theory are true. We must of course understand the notion of "truth" here in the way we have taken it above. We could say that mathematical intuition gives evidence for the truth of the axioms. This might be viewed as a kind of intuitive soundness proof, or at least as an explication of what it would mean to have an intuitive soundness proof. For if we take individual variables to range over natural numbers and axioms for identity to be interpreted in accordance with our remarks on the construction process, the notion of a successor x' of a number x is evident and so are the properties that $x' \neq 0$, and that $x = y \leftrightarrow x' = y'$. The induction principle which states that for S a property of the natural numbers,

$$(S(0) \wedge (\forall x)(Sx \rightarrow Sx')) \rightarrow (\forall x)Sx$$

is also evident.

We cannot of course assume uncritically that mathematical intuition provides the absolute truth about numbers, or that it gives unassailable answers to our questions about numbers. We have not tried to claim that the evidence it gives is necessarily adequate or apodictic, but we have pointed out some of the features of the evidence that are relevant to its "objectivity". What we have been considering are the "categorial activities" involved in intuition which produce beliefs we have about numbers. These processes provide the relevant evidence for our beliefs about numbers. By focusing on the role of these mental processes in our discussion of arithmetical knowledge we have avoided quandaries about mathematical knowledge and causality. In order to understand constructions and their properties it would be quite out of place to bring up questions about causality. The investigation of constructions surely requires no analysis of causal relations. This is just another way of looking at our point above about the phenomenological reduction. In making the phenomenological reduction we are simply to investigate the structure of acts and sequences of acts and to do so does not require considerations about causality. We might say, in a way that is reminiscent of some of Gödel's remarks, that the knowledge we have of numbers by way of intuition is analogous to the knowledge we have of objects of the physical world by way of straightforward intuition. On the basis of our remarks on evidence we should have no more worry about deception in arithmetic than we do about deception in sense experience. This is not to say that such deception does not occur. In fact we shall have more to say about deception in mathematical experience in the next chapter. The point to be made, however, is that this just supports our general thesis that mathematical and perceptual intuition are analogous in several important respects.

It should be noted that the fact that conclusions by induction can be taken to give intuitive knowledge suggests a philosophical viewpoint that goes beyond Hilbert's original conception of finitism, or beyond a conception of finitism that would identify the mathematical content of the finitist viewpoint with primitive recursive arithmetic. Where our view differs is in recognizing the role of reflection in seeing that induction holds as a general principle, and in the further supposition that what is known by reflection must itself by produced by intuitive processes. It is in this latter use of intuition that we could be said to have evidence for the truth of general arithmetical state-

ments. For what reflection gives us is what Parsons once described in relation to intuitionism: "Intuitionism requires that our insight into meanings or intentions be sufficient to not only establish laws directly relating to objects in space and time but also to establish laws concerning the *intentions* or mental constructions".[6] What has not often been noted is that the process of reflection must itself have an intuitive character. To defer to reflection is not therefore to drop the notion of intuition in an account of mathematical knowledge. Further investigation of the notion of reflection might be one way in which we could begin to make a philosophical case for the claim that constructive principles that go beyond "concrete mathematical intuition" of sign configurations nonetheless do have an intuitive foundation, a claim that has been suggested by several logicians as mentioned at the outset of this essay. One might naturally suppose, for example, that processes which operate on objects like natural numbers can themselves be taken as objects upon which other processes operate, much as one thinks of this as happening in certain hierarchies of recursive functions.

CHAPTER 7

FINITE SETS

1. INTRODUCTION

A theory of finite sets can be formulated in such a way that it is quite analogous to arithmetic. Hence much of what we already said about natural numbers will have application in this chapter. In this chapter, however, we shall treat some questions and issues that are specific to constitution of the consciousness of finite sets. Questions about the intuition of finite sets are of interest in their own right but this chapter can also be viewed as providing part of an account of the notion of "multitudes" as objects that are intuited in Husserl's conception of numbers as "determinate multitudes". In *PA* Husserl tended to treat the concepts of "multitude" and "set" as synonymous, claiming that the concepts were undefinable. Hence, what is needed philosophically, as in the case of any of the primitive concepts of mathematics, is an analysis of their origins. In this chapter we shall be speaking only about finite multitudes, and we shall again be applying the Kantian strategy. Thus we shall be claiming that there is knowledge of finite sets, that we do make determinations about finite sets, and then asking how this is possible. An analysis of origins provides the answer.

In *PA* Husserl was interested in the conception of numbers as determinate multitudes. Of course he did not have in mind, as we have seen, the kinds of reductive definitions offered by Frege and others. On the basis of such definitions it is clear that we can view numbers as finite sets, for we can clearly model number theory in (finite) set theory in a variety of ways, as on the von Neumann or Zermelo construals. But of course we can also go the other way around and view finite sets as numbers by means of suitable encodings of these objects. It is then possible to model finite set theory in number theory, as in Ackermann's model. The usual set theories, however, are much stronger than number theory and could not be modelled in it. If we were to start with number theory we could build up only a part of set theory, what might be loosely thought of as the combinatorial part. There would be aspects of the more general concept of set that we would not be able to touch. One such aspect that might be of interest phenomenologically would be what Bernays called the "quasi-combinatorial" character of the way in which

transfinite sets are built up, according to the iterative conception of set. But as we mentioned, we shall not attempt to treat questions about transfinite sets here. Instead, we shall consider some elementary questions about the origin of the set concept. For even reductive definitions of number in terms of sets presuppose the concept of set as a primitive.

From our phenomenological point of view we shall regard finite sets as objects obtained from given objects (initially individuals) as a result of a process of construction. Intentions directed to sets may be thought of as referring not only to objects thought to exist independently of us but also to processes for obtaining these objects. The point of phenomenology is then to analyze (empty) intentions directed to sets as expectations or anticipations about the course of our experience with respect to these objects. In particular, we would want to know whether and in what way the experience is well-determined. *Intuiting* a finite set amounts to actually carrying out the acts specified by the process or procedure in which these objects would be given. In that case we actually see what was merely anticipated or thought about. The intention directed to a particular set might be fulfilled, partially fulfilled, and so on. Then we can regard set-theoretic knowledge as a product of intentions and intuition. In this chapter we shall examine elements of the construction process in some detail.

The intuitive processes involved in providing evidence for M's belief in the *existence* of finite sets will be of special interest. We shall start with the processes involved in fulfillment of intentions directed to quite small finite sets, and in singular statements about them. The notion of the horizon of acts directed to these objects is again relevant. The horizon of such acts is to be regarded as determined by precisely what the intention of the act is and by our background beliefs about the object or objects in question (in our case HF sets). We shall also discuss the role of founded acts – acts of reflection and abstraction – in providing evidence for these objects. Some comments will be made about the evidence given by the construction process in the case of quite small finite sets. Set-theoretic operations and relations involved in singular statements will be considered and some difficulties about the construction of the empty set and unit sets will be addressed. It would already be useful in this context to briefly compare our view with some ideas of Gödel and Wang on "running through" or "collecting" objects, on the function of "synthesis", and on sets as "wholes" comprised of "parts".

The intuition of large finite sets will then be considered. In the case of fulfillment of intentions to large sets the horizon determined by the act's intention will be important. For the intention could actually be only partially

fulfilled and a condition for regarding it as fulfillable would be that the process be well-determined. We cannot actually complete the "synthesis" of the whole but we would want to know how it would be possible. The idea that intuition of a set as a whole is founded on intuition of its parts will play an important role in this context. After considering in a general way the construction of large finite sets we shall focus on *iteration* of the process of constructing finite sets and, on the analogy of our remarks in Chapter 6, the question of what is involved in seeing that it is *possible* to continue the process of constructing finite sets. The evidence provided by the acts that produce M's belief that it is possible to continue the construction process for finite sets will be considered. These acts must again be acts of reflection that have the construction process as their object. In particular, the acts involved in obtaining a hierarchy of finite sets, the HF sets, and the structure of this hierarchy will be considered. It is pointed out how the hierarchy just depends on the structure of founding and founded acts in which sets higher in the hierarchy must be constructed from objects already given at earlier stages in the process, and ultimately from individuals given in straightforward perception (on a Husserlian view). In this context we shall briefly discuss the function analogous to "representing content" in acts of set-theoretic intuition and some other philosophical issues about set-theoretic knowledge.

In conclusion we consider how on our account set-theoretic intuition may be viewed as giving evidence for the truth of the axioms of finite set theory. We do not make the uncritical assumption that it gives knowledge of truth in an absolute sense. It will be argued that reflection on the construction process for HF sets provides evidence for general statements about finite sets.

2. A THEORY OF FINITE SETS

A theory of finite sets, looked at in the proper way, will be quite analogous to arithmetic. We shall consider a theory of hereditarily finite sets with individuals.[1] The natural numbers are obtained by beginning with 1 (or with 0) and iterating the process of constructing a unit an arbitrary finite number of times. Similarly, for a given domain of individuals D the HF sets built up from D, HF(D), are obtained by beginning with the empty set Ø and iterating the process of constructing from any $x_1,\ldots,x_n$, individuals or previously obtained HF sets, the set $\{x_1,\ldots,x_n\}$ whose members are just $x_1,\ldots,x_n$. For any D, D∪HF(D) forms a model for the usual ZF axioms of set theory without the axiom of infinity. The n-adic operation of forming a set from an arbitrary finite number of given elements could be replaced by the dyadic operation of

forming from a previously obtained set x and an object y the set $x \dotplus y$ whose elements are just those of x, together with y, i.e, $\{x_1,\ldots,x_n\} = (\ldots(0 \dotplus x_1) \dotplus \ldots \dotplus x_n)$. Then we can think of D∪HF(D) as obtained by beginning with Ø and elements of D and iterating the process of forming $x \dotplus y$ from previously given x and y, i.e., of adding a new element to a previously given set. Looked at in this way an arithmetic-like theory of HF sets based on Ø and $\dotplus$ is straightforward to formulate. Corresponding to the elementary Peano axioms would be comprehension axioms for Ø and $x \dotplus y$, and extensionality (writing 'Mx' for 'x is a set', i.e., '$x = \emptyset \vee (\exists z)(z \in x)$'):

$$x \notin \emptyset,$$
$$z \in x \dotplus y \leftrightarrow (z \in x \vee z = y),$$
$$Mx \wedge My \wedge (\forall z)((z \in x \leftrightarrow z \in y) \rightarrow x = y).$$

Then we have the induction axiom:

$$(F\emptyset \wedge (\forall x)(\neg Mx \rightarrow Fx) \wedge (\forall x)(\forall y)(Fx \rightarrow F(x \dotplus y))) \rightarrow (\forall x)Fx.$$

With a schema analogous to the primitive recursion schema for introducing functions in first-order arithmetic we can derive the ZF axioms without infinity.[2] It will be useful to think of the HF sets in terms of this formulation in the remarks that follow since it captures in a natural way a number of remarks that have already been made in connection with the natural numbers.

One remark that we should make about the identity of sets is that in general we should be prepared to recognize a difference between extensional and intensional identity. Extensional identity is just given in the usual way by the axiom of extensionality. But in general since sets will be initially specified in comprehension axioms by intentions, two sets could turn out to be extensionally identical even when they are given by different definitions or procedures. Intensional identity would hold when there is an identity of intentions directed toward sets. The distinction between extensional and intensional identity is more important for infinite sets than it is for finite sets since we should evidently not separate the output of the process of constructing an infinite object from the process itself. We could not regard the process as completeable in the same sense as we could for finite sets. In the case of infinite sets the extension will never be completely formed. Of course in the case of large enough finite sets one might argue that the same is true, but at least our idea of completeability in this case does not abstract from some of the basic conditions on intuitability for M. In any case it should be recognized that the necessity of distinguishing between intensional and extensional

identity will be sensitive to these kinds of issues.

We shall think of the intuition of finite sets as founded on the intuition of their elements. Without intuition of at least some of the elements there could not be intuition of the set. A set will be viewed as a certain kind of "whole" which is intuited on the basis of "synthesis" of its parts. Once we have developed a number of observations about this in the case of quite small finite sets it will be possible to discuss the general view that if D consists of possible objects of intuition, then HF(D) will be constructible in intuition. Since the fundamental conception of set that will emerge is one according to which a collection of previously given objects may be regarded as forming or constituting a single "totality" or set our view will yield what has been called a "mathematical" conception of set. The "mathematical" conception has been distinguished in the literature from the "logical" conception of set according to which we are to think of sets as given by extensions of properties or concepts where the entire universe is separated into two categories depending on whether or not things have the given property or fall under the given concept.[3] The mathematical conception is appealed to in iterative conceptions of set, even in the case of what has been called the "maximal" iterative conception. For even in this case we have what Bernays called a "quasi-combinatorial" notion of set, not a "logical" notion.[4]

3. THE ORIGIN OF THE CONCEPT OF FINITE SET

In order to begin our consideration of the process involved in fulfillment of intentions to quite small finite sets we should consider once again the retentive-protentive structure of consciousness upon which the intuition of sets is founded according to Husserl's views. We shall consider how this structure must be involved in constitution of the consciousness of sets. In Chapter 5 we already discussed some of the features of this structure of acts that are most relevant. We should now consider again the structure as we depicted it in Figures 1 and 2.

The main idea is that given some objects denoted by a, b, c, d, e, we can "collect" these objects by running through them one by one. In the most primitive setting, before (founded) acts of reflection and formal abstraction are involved, the objects would just be ordinary kinds of individuals given in straightforward perception, e.g., bottles, matches, even symbol tokens like our stroke inscriptions. The awareness of sets is not equivalent to but is founded on seeing them as a group or *as a collection.* Of course in seeing individuals as a collection nothing about the individuals changes. What

changes is our *awareness* of them, the way we regard them. In the case of very small collections we can see the individuals as a collection in one act. It is not necessary to undertake a sequence of acts in order to regard them as a collection. But if we are interested in the relevance of the phenomenon to mathematics this will prove to be the exception rather than the rule. In the process of collecting it is our awareness of the individuals that changes and the question of just how it changes is what we now take up in more detail.

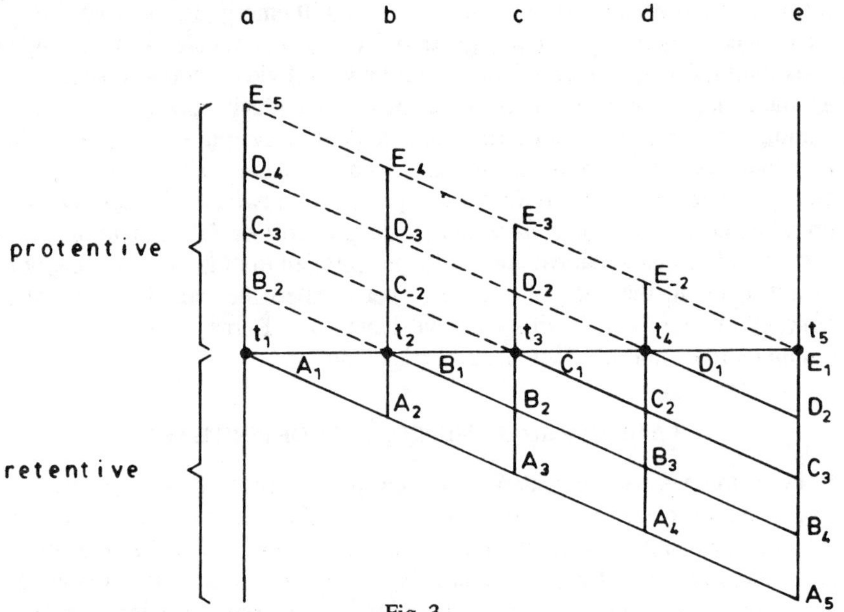

Fig. 3.

Recall that in our discussion of the structure depicted in Figure 3 we said that at each stage in the process we take another item into account and it thereby becomes unified or synthesized into a whole in one act at that stage. The act at that stage of course contains what was already seen, as indicated by the vertical lines representing the retentive structure at each stage. We can continue to collect items in this way so that the items are taken into consciousness one by one. Those items which are not taken into account at a certain stage but which may be protended, or may be in the background (i.e., in the external horizon of the act at that stage) are not unified in one act and hence

are not part of the collection. As we said earlier, there is a fundamental difference between our awareness of the items that have been explicitly seen and those that have not. The items that have been taken into account at some stage are synthesized into a whole in *one* act at that stage but the other items are not yet unified in one act. They will become so as we continue to run through the group and this will happen automatically by virtue of the act-structure involved. So at each stage we have *one* act which contains the *many* items that were seen at the previous stages. This is the kind of phenomenon that from a phenomenological point of view underlies Cantor's view of a set as a "many (Viele) which can be thought of as a one (Eines)". For if at a given stage we did not have *one* act which contained the *many* items seen at previous stages we would not have conditions necessary for fulfillment of the intention to a *set*.

As Husserl describes the matter, what is required for intuition or construction of a set is not that we just continue to run through items one by one, but rather that we regard or look at the collection which results from this process *as a single object*. To see it as a single object we must have very basic conditions for individuating it, for taking *it* as an object. These conditions are constituted by the process of actually running through the items that are given, for there is a fundamental difference in this process between the items that have been synthesized or collected into a whole and those items which are not part of this unity. So it is possible to regard or look at the collection as a single object if we become aware of the difference between it and what is other. A condition necessary for intuition of the set as an object is that we be able to distinguish what is in the collection from what is not in it. To become aware of this difference itself involves *an act* – an act directed toward what is constituted in the process of running through the given objects. This further act is an act of reflection. In his description in *EJ* Husserl refers to it as a "retrospective apprehension" (ruckgreifendes Erfassen). Such an act of reflection is necessary if we are to regard a collection of previously given objects as a single object, a set. Of course this by itself does not give intuition of a set as an abstract object. Acts of abstraction must also be involved, for it is the structure of collecting which remains the same across cases of collecting different kinds of objects.

Husserl's own description in *EJ* is worth considering for the contrast it provides between the founding acts involved in the intuition of sets and the founded acts in which sets are given. He says there that

> ... in order for the collective connection, originally sprung from the act of plural explication...to become a substrate – i.e., a true object, something identifiable – a *turning of regard* is first required. But this implies that as long as we carry out a merely collective assemblage we have more than ever only a *preconstituted object*, a '*plurality*', and only in retrospective apprehension, following the active constitution, do we have *as an object plurality as unity: as set*...The collective synthesis, the "A and B and C" is indeed the noetic unity of a consciousness but not yet the unity of an object in the proper sense, that is, in the sense of a thematic object-substrate. Here A, then B, then C is thematic, but the collective is not yet thematic. The colligating (kolligierende) consciousness contains several objects encompassed in unity but not a unique object having several members. Nevertheless, through every synthetically unified consciousness a new object is essentially preconstituted, precisely one having many members; there is then required only an act of thematic apprehension, possible at any time, to make what is thus preconstituted into an object and thereby a substrate of judgment... Subsequently, it is an object like any other; not only can it be totally identified as the identical element of many modes of giveness, but it can be explicated in an ever renewed identification; and this act of explication in its turn is always an act of colligation....[5]

As Husserl points out in this passage, the collective synthesis is indeed a "noetic unity of consciousness" but it can become an *object* only through an act of reflection in which this noetic unity is itself individuated. Reflection is required to unify or synthesize in one act what is given in the different acts that are carried out successively in time. In the case of very small sets this would not be required since we can already regard the elements as forming a set in one act, without having to successively run through them. But this would be the exception rather than the rule. We might note that the type of synthesis brought about by what we are calling reflection is much like what Kant refers to as "the synthesis of recognition in a concept" in the transcendental deduction of the first edition of the *Critique*.

To intuit a set thus requires an act of reflection. In this act of reflection the collection of previously given objects is regarded as a single object. This means, in effect, that it becomes the "x" of an act of reflection. From the point of view of the act structure it is what is being taken as the "determinable x" of the act of reflection. The conditions for individuating a particular set are constituted in the underlying process of running through

some items in which at a given stage some of the items are contained in one act by virtue of the retentive structure of consciousness. If we adopted a notation which distinguished variables to be used only for sets, say, upper case letters X, Y, Z,..., from variables to be used for either sets or individuals, lower case x, y, z,..., we could think of the determinable x of acts of reflection of this type as expressed in a formal theory by the upper case letters. The x of an act of reflection is then of course an object for further predicates and these predicates may be either true or false of it (they may either be fulfilled with respect to it or not). In some of his works Husserl describes this as the process that underlies "nominalization".

We might picture this process in its most elementary form in another diagram. Suppose that we are given the individuals a, b, c, d, and e, and that we run through the individuals a, b, and c in the order indicated by this diagram. We can represent the construction of the set {a, b, c} in the following way:[6]

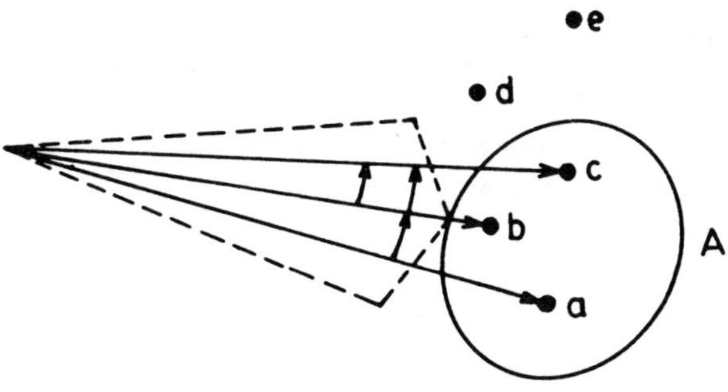

Fig. 4.

The unbroken lines to a, b, c represent the acts directed toward these objects as we run through them one by one. The vertical arrows running between the unbroken lines represent the retentional structure involved as we run through the objects, beginning with a. The diffuse arrow formed from broken lines represents the act of reflection on the collection constituted by running through the items in which we have the construction of the set {a, b, c}. Note that we shall understand the bracket notation for sets as simply expressing that this process has been carried out (for large sets $F = \{x \geq \phi\ x\}$, that it could be). We take the brackets in the example to be an expression of the fact that an act of reflection has been performed, and then on this basis, an act of

formal abstraction, which has as its object the particular collection constituted by running through some objects, where we leave out considerations about order. Of course we need not leave out considerations about order. We could express them by usual notational conventions for ordered sets. It should also be clear from the discussion thus far how we are interpreting ' ∈ '. More will be said about this in a moment.

It should be noted that in our diagram the individuals e and d do not belong to the set. We could of course construct a set in which they would be contained. In fact, we could construct all of the different possible sets of these objects. To each of these particular construction-objects would correspond a different construction-process. The processes involved in the construction of finite sets from given individuals could be readily represented in tree structures, as might be suggested by Figure 4. The nodes in trees would represent results of reflection on particular collections.[7] If we allow infinite branching then every element of the universe of sets built up from D can be represented by a tree, so that what is expressed by bracketing could be equivalently expressed by trees. Taking the branches to be unordered the representation is unique. ϕ is represented by a one-node tree with nothing attached at the node. $x \in D$ is represented by a one-node tree with x attached. If y is a non-empty set it is represented by a tree with y at the vertex, with a successor node corresponding to each element x of y, with the tree corresponding to x attached:

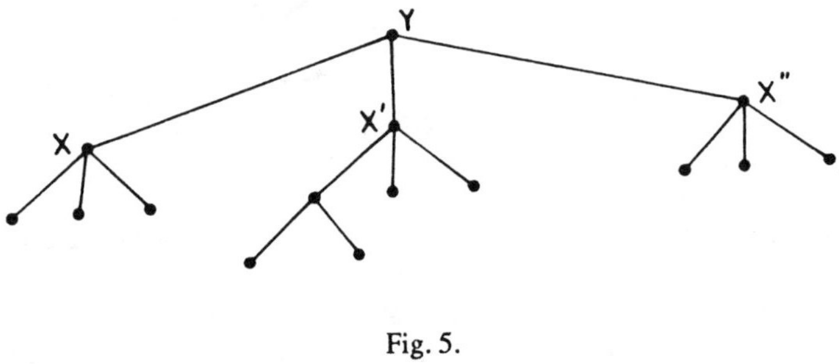

Fig. 5.

The end nodes correspond to the ultimate elements of y (i.e., the individuals, including Ø, in the transitive closure of y). The construction-object at a node in the tree is determined by the construction-objects at successor nodes, except in the case of the end nodes. We discuss this in more detail below,

including the well-foundedness of the sets.

4. INTUITION OF FINITE SETS

On our view, therefore, the intuition of finite sets takes place in founded acts of the type that Husserl often refers to as "higher-order" acts. As we said from the outset, these higher-order acts depend on the fact that certain other acts have already been carried out or lived through. Once these acts have been carried out they can become the objects of further acts. Acts of reflection and abstraction are object-directed acts just like any other intentional acts. They must be viewed as processes that produce certain beliefs, or that give evidence for certain beliefs. For certainly some beliefs that we have about acts or about the structure of mental processes are more reasonable than others, so that we have evidence for some of these beliefs. Reflection and abstraction thus have an important role in set-theoretic knowledge because just like any other acts, acts of reflection and abstraction can be fulfilled, partially fulfilled, frustrated, and so on. These acts are characterized by the difference between mere intention and intuition. The intention to a particular set will be fulfilled when we can carry out the appropriate underlying acts and the acts of reflection and abstraction in which the resulting collective synthesis can be regarded as a single object, a set.

The fundamental phenomenon from which the intuition of sets is built up is that of seeing some ordinary individuals given in straightforward perception as a group. Straightforward perception provides the immediate, concrete foundation from which intuition of mathematical objects must be constituted. The intuition of sets is founded on this phenomenon when the structure or form of the given underlying process becomes the object of our act of abstraction. For much as we noted before that the concept of number is more general so that any objects can be numbered, we should now note that the concept of set is more general so that any items, not just these particular perceptual individuals, can be collected into a set. At a first level of abstraction it might be supposed that it is not just the particular given perceptual individuals with their particular qualities and space-time locations that can be taken as the elements of a set, but that other kinds of perceptual objects with different space-time locations might also be taken as elements of a set. At a higher level of abstraction we need not suppose that the individuals have to be only of the kind of perceptual individuals. Once we have abstracted from the phenomenon of seeing *perceptual* individuals as a group it is clear that we could take the individuals to be any objects that we could regard as intuitable,

including, for example, mathematical objects like integers. The individuals would not have to be construed as concrete objects located in space-time.

As we said earlier, acts of formal abstraction are to be understood as founded acts in which we attempt to determine the form or structure of what is given. If we were not concerned with the structure of the underlying process we could not suppose that the intention to a set as a mathematical structure was being fulfilled. What is relevant to the set as a mathematical structure is carried along in our notation. It should be noted that we are saying that in acts of abstraction it is our awareness of the individuals that changes, the way we regard them, not the individuals themselves. We regard the particular perceptual individuals simply as particular instances of the determinable x's of the underlying acts of straightforward perception, where it is clear that these x's could be any particular individuals at all. The x's in the underlying acts could have a different content depending on the context. In fact in *applications* of set theory in which individuals would be called for that is just how we would think of the sets. The sets in a given context would be sets of particular objects. In applications the qualitative features and possibly space-time locations given in the underlying acts would be relevant to fulfillment of descriptions of the sets involved.

A consequence of Husserl's views about founding is that at the earliest stage in the construction of sets the sets will be comprised of individuals and only later in the process of constructing sets will sets be included among the objects that can themselves be collected. In *EJ*, for example, Husserl says

> But finally every set, preconstituted in intuition, leads to ultimate constituents, to particularities which are no longer sets. For it belongs to the idea of such a set that in its first giveness as substrate there is already present a pregiven multiplicity of particular affections which we actualize by its apprehension...*every set must be conceived a priori as capable of being reduced to ultimate constituents,* therefore to constituents *which are themselves no longer sets.*[8]

Sets are thus to be thought of as well-founded. This view, I would argue, is essential to the Husserlian conception of the role of founding and founded acts in our knowledge of abstract objects. For it is claimed that such knowledge must ultimately be founded on straightforward perception. Straightforward perception provides the immediate, concrete, pre-reflective basis of our experience. Objects like sets are not given in straightforward perception: we do not literally see sets in straightforward perception. What we do see are ordinary perceptual objects given in space and time, charac-

terized by properties befitting such objects. Abstracting from this basis we obtain a formal notion of individuals. This already involves founded acts, for we are then seeing that the particular individuals involved in some collection are irrelevant to regarding the collection as a set. There can be intuition of sets only in founded acts, for the way we regard the originally given individuals changes. It is no longer these perceptual individuals and their particular properties that are of interest. This change indicates a difference in the thetic character of the acts involved. In acts of formal abstraction and reflection we are no longer engaged in the simple, straightforward intuition of perceptual objects. The sets obtained at the earliest stage in the hierarchy of founded and founding acts will therefore be the sets constructed from individuals, whatever the individuals may be. We might then take the sets obtained at this stage to themselves be objects that could be collected. We shall discuss this latter possibility in more detail below.

It can again be argued that there is in acts of set-theoretic intuition a function analogous to that of representing content in ordinary perceptual intuition. The acts will be constrained and determined by what is "given", just as we suppose this is so in ordinary perception. In the construction process for sets the founded acts are determined by just what is given in the founding acts. Where individuals are involved the process of collecting will be largely individuated by the objects collected. The relevant "given" for set-theoretic intuition will consist of the particular underlying process that is carried out with respect to the individuals. The structural features of the construction process will of course still be relevant, but identity conditions on impure sets will be different from identity conditions on pure sets. In the case of pure sets the analog of the representing content will be appropriate to abstract objects, for sensory data will be irrelevant to the structural features of the process. In the case of impure sets, qualitative features associated with the acts involved may be relevant to individuation, so that impure sets will not be individuated solely on the basis of their structural properties. But in either case we determine what is given in the founded acts on the basis of what is given in the founding acts, and what is given in the founding acts cannot be altered at will. The behavior of the construction process with respect to the objects in either case determines and places constraints on the intuition. For example, the particular construction process in which we run through the individuals denoted by a, b, and c, for example, cannot be altered in such a way that we could regard it as the particular construction process involved in running through the individuals denoted by a, b, c, and d. We could think of these determinations as being made on the basis of the underlying in-

dividuals, in addition to the process of giving 1-to-1 correspondences. It follows that because the act is determined on the basis of the given underlying process it would be nonplastic in the way required for supposing that we have intuition of, not mere conception of, an object.

Since the acts are fixed on the basis of what is previously given there can be error in set-theoretic intuition just as there can be in straightforward intuition. One way that an (empty) intention to a particular set will be frustrated is in the case where in the process of its fulfillment we are given an object different from the intended object, i.e., where on the basis of underlying acts we do not arrive at the object intended. In straightforward intuition an intention to a particular object will also be frustrated when in the process of its fulfillment we are given an object different from the intended object. Of course in the latter case this will happen because of various sensory determinations which fix an object as different from the object intended. Sensory determinations will not play the same role in set-theoretic error. For set-theoretic intuition the relevant content is specified by the particular process of collecting that has been undertaken. The "content" of the reflective act in which a set is individuated is the collection constituted by a particular act of running through some given objects. This may or may not match the intention.

5. COMPARISON WITH GÖDEL AND WANG

These considerations might be briefly compared with ideas of Gödel and Wang on sets and set-theoretic intuition where we think of their ideas as applying to finite sets.[9] For in remarking that set-theoretic intuition is analogous to perceptual intuition Gödel suggested that in perceptual intuition we may view objects as formed on the basis of synthesis of the different parts of the objects that are given to us. In the intuition of sets we have something similar: a set is formed on the basis of synthesis of different parts of it which are given to us. Of course in perception sensory information plays a role that it does not play in set-theoretic intuition. Gödel says of this fact that mathematical intuition need not be viewed as a faculty giving immediate knowledge of the objects concerned, but rather that, as in the case of physical experience, we form our ideas of objects on the basis of something which is immediately given. But in the case of set-theoretic intuition what is immediately given is not, or is not primarily, sensations. Gödel surmises that the "given" underlying mathematics is "closely related to the abstract elements contained in our empirical ideas".

These views could evidently be interpreted in our account in a fairly direct way. There are several points in the quotations from Gödel in Chapter 1 of which we should take note: (1) objects, including sets, are formed on the basis of *synthesis* of different parts of the objects which are given to us. (Perhaps Gödel should have said that *knowledge* or awareness of objects is formed on the basis of synthesis); (2) intuition of sets need not be conceived as giving *immediate* knowledge; (3) rather, we *form* our ideas of objects, including sets, on the basis of something which is immediately given; (4) the immediately given in the case of sets is not, or is not primarily, sensations; (5) even in ordinary perception something besides sensation is given; (6) there is an analogy between mathematical and physical intuition insofar as mathematical intuition induces us to build up mathematical theories, just as physical intuition induces us to build up physical theories, and, in both cases, intuition induces us to *expect* that future intuitions will agree with those theories, or with past intuitions; (7) Gödel also mentions the analogy in each case of believing that questions not decidable now have meaning and may be decided in the future.

Concerning (1), Gödel's remarks can be directly interpreted in our account. Gödel remarks in his paper that the function of the concept of set he discusses is "synthesis" in a sense close to Kant's. Synthesis, we might say, is a process of bringing separate intuitions together into a whole. If we followed either Kant or Husserl we would of course want to distinguish between different kinds of synthesis, including active and passive synthesis. But the basic idea underlying Gödel's remark is that as a result of the function of synthesis in ordinary perception we intuit *one* object on the basis of various aspects of it which are given to us, and that the same will be true in the case of set-theoretic intuition. Gödel's point in the text about synthesis in the case of sets however is that, unlike Kant, we should regard this kind of synthesis as also providing "objects"; that is, it does not follow that what is given on the basis of synthesis in set theory should be regarded as purely subjective, as Kant thought, just because the "data" involved in set-theoretic intuition cannot be immediately associated with the action of certain things upon our sense organs. This point, it seems to me, is very much in the spirit of Husserl's entire conception of categorial objects and categorial intuition, and it would certainly be a point on which Husserl would claim to have improved upon Kantian epistemology.

(2) can also be directly interpreted in our account. For, on our account, set-theoretic intuition could not be presumed to give immediate knowledge of the objects involved. In fact the notion of immediacy is used in phenomenology

as a criterion for distinguishing straightforward from categorial intuition. Set-theoretic intuition gives mediate knowledge, founded on acts that have already been carried out. This leads directly to point (3), for on a phenomenological conception an act of intuition, whether it be perceptual or mathematical, will have the characteristic that the "matter", which is understood as the "given", will be interpreted in one way or another in the act. Husserl adopts a form/matter distinction of just the type that Gödel is invoking. But then, by point (4), the given (the matter, or "representing content") in the case of sets is not, or is not primarily, sensations. What is relevant about sequences of founding acts, if we are concerned with fulfillment of *mathematical* intentions, is their structure, not the sensory data that they may contain. Founded acts of reflection and abstraction are required for set-theoretic intuition. We might attempt to interpret Gödel's remark that the "given" in mathematics is "closely related to the abstract elements contained in our empirical ideas" along the lines of ideas about acts of abstraction discussed above. For then we could say that the objects that are given and out of which we construct sets can be any objects at all. We regard them as particular determinations of the "determinable x" of each act involved in collecting. Just what kinds of qualities they have, as determined by various sensory elements, will not be relevant to the construction of the set.

(5) can be readily interpreted in our remarks of Chapter 3. Perception has a form/matter structure, the same sensory data can be interpreted in different ways, and so on. We have also discussed the nature of expectations about intuitions in some detail so that we can readily interpret (6). And we have already commented on the notion of meaning in such a way that (7) would be true.

Finally, we might note that on our account causal conditions would not be involved in the analysis of evidence provided by set-theoretic intuition even in the case of quite small finite sets. This feature of the phenomenological view would no doubt also be of interest in developing ideas like Gödel's.

Gödel's remarks on set-theoretic *intuition*, like Wang's, strongly suggest that we interpret the idea of set formation and other operations and relations in set theory in terms of operations of the mind, much as we are interpreting them here. Fortunately we need not address the very difficult questions that can be raised for the attempt to extend this interpretation to the kinds of sets that can be shown to exist in the presence of axioms of infinity, replacement, and power set. It is in the context of full ZF that such an interpretation would be especially difficult to reconcile with the set-theoretic realism suggested by other remarks of Gödel.

One interpretation of Gödel's ideas which purports to show that set-theoretic intuition is reconcilable with set-theoretic realism is due to Maddy.[10] Maddy is interested in particular in showing, in response to the kind of problem posed by Benacerraf, that a causal theory of knowledge for sets is reconcilable with set-theoretic realism, where we think of such realism as a consequence of the "standard" (Tarskian) theory of truth for the language of set theory. On a causal theory of knowledge of sets there would need to be a causal chain of reference going back to an initial "dubbing" or "baptism" of such objects and to establish the plausibility of this Maddy argues that we literally perceive *sets* of physical objects with our senses. She first argues that our perception of physical objects in general depends on the development of neurophysiological "object-detectors". Interpreting Gödel's remarks she says that the presence of a physical object in our physical experience is due to our neural object-detectors. The object-detectors are the "abstract elements in our experience" which are "due to another kind of relationship between ourselves and reality", that is, a relationship other than "the actions of certain things upon our sense organs". On her view the action of a physical object on our sense organs is responsible only for our sensations or the pattern of sensory stimulations, but the object-detector is responsible for the idea of the object itself. The presence in us of the object-detector is a result of a much more complex interaction between us and our environment than that which produces sensations. This complex interaction is nonetheless causal. It is what makes up Gödel's "other relationship" according to Maddy. The object-detector is what legitimizes the gap between what is known about and what is actually causally interacted with.

What Maddy has to say about perception of sets is developed on the analogy of her suggestions about perception of physical objects. Her view is that we perceive sets of physical objects by virtue of the development in us of a neural "set-detector". The set-detector results from interactions with sets of physical objects which bring about structural changes in the brain by some complex processes of a type suggested by certain neurophysiological studies. Thus, when we look into an egg carton, for example, and see three eggs we have acquired some numerical beliefs about the eggs. These beliefs, Maddy argues, are perceptual. She claims further that they are perceptual beliefs about a *set*, namely the belief that there is a set of eggs before me, that it is a three-membered set, and that it has various two-membered subsets. Should the belief that there are a number of eggs be regarded as a belief that there is a set of eggs having that number? On the face of it these would seem to be different beliefs, for we could easily imagine ordinary perceivers who believe

they are seeing a number of objects without believing that they are seeing a set of objects. The belief that they are seeing a set of objects would seem to require some further knowledge. In ordinary experience we often see a number of objects without thinking of the objects as forming a set. In any case this may not be essential to Maddy's account since the underlying phenomenon about which she is concerned is probably just that of seeing physical objects not only individually but also in groups. Seeing physical objects in groups would presumably already involve a set-detector. I am skeptical however about assimilating perception of groups of physical objects to perception of sets for reasons that Maddy mentions as grounds for objection to her account. I shall consider this objection in a moment.

We might first ask how faithful this interpretation is to Gödel's ideas. As speculation about the neurophysiological basis of ordinary perception of individuals or of small groups of individuals it is perhaps plausible but it is not clear, at least on certain points, that it would coincide with Gödel's views. For example, Gödel states that "another kind of relationship" to reality is involved in intuition of sets, by which he may have in mind not only that the relationship is non-immediate but also that it is non-causal. The impetus for talking about a "relationship" that is non-causal would evidently come from the fact that he is speaking of the truth of statements about the objects of transfinite set theory which "clearly do not belong to the physical world". Moreover, he only says that there is "something" like a perception" of transfinite sets, certainly not that we literally perceive them. Maddy's paper does not speak to the question of perception of transfinite sets but it is especially in this context that one might be inclined to loosen or even drop causal conditions on knowledge in order to maintain set-theoretic realism. One might then speak of other kinds of "relationships" to these objects, e.g., as phenomenologists speak of intentional or phenomenological "relationships" to objects. In that case the notion of a "relationship" gets interpreted in a very different way. I do not know that Gödel had this in mind, but I do not think that it, or a view like it, could be rejected out of hand.

In any case straightforward perception of small groups of individuals cannot be so readily assimilated to perception of sets. Maddy notes that it is an objection to her account that a person could not be perceiving a set in the egg carton before him because sets are not the kinds of things that have location in space and time. Maddy admits that this is so for many sets, e.g., the empty set, the set of real numbers, but she believes that sets of physical objects do have location in time and also in space insofar as their elements do. She wishes to deny that abstract objects cannot exist in space and time,

for sets of physical objects do so exist. This view seems to me to be confused. As Chihara has pointed out, if we consider the perception of the egg on a table then according to Maddy's view we literally perceive not only the egg but also the set which happens to consist of just this egg, i.e., a unit set.[11] But then we should also be able to perceive the set which has as its sole member this unit set. This set is located at exactly the same place and time as the egg. Moreover we should be able to perceive the set consisting of the egg and the unit set which has as its only member the egg. In fact, there would be an indefinite number of sets that we should be able to perceive in exactly the space-time location of the egg.

Maddy's set of three eggs is supposed to exist at the exact place where the three eggs are seen. It presumably came into existence when they did, and will cease to exist when they do. The set could also presumably be moved about in space-time. Now it seems fine to assert that these things are true of the individuals that we perceive and perhaps also of groups or aggregates of physical individuals, provided we think of the group or aggregate as individuated in some way by physical properties. Then we could say that group of physical objects does have location. But the properties of groups of physical objects would then surely be different from properties of sets unless of course the notion of "set" is not being used in a mathematical sense at all. The problem with a view like Maddy's is that if we insist on literal perception, or what we have called "straightforward" perception, we do not get objects that could count as genuine sets, as objects that a mathematician would recognize as sets. For sets are not given in straightforward perception even though we might suppose that under certain circumstances the elements of sets are. If we *do* get genuine sets then we are not really insisting on straightforward perception. On our view sets are not of course literally perceived but rather there can be intuition of sets only in founded acts.

6. UNIT SETS, THE EMPTY SET, AND MEREOLOGY VS. SET THEORY

We should now consider several questions that might be raised about the intuition of unit sets and the empty set, and also about the appropriateness of thinking of sets as "wholes" formed from various "parts". We can then briefly discuss how set-theoretic operations and relations are to be understood. The distinction between intuition *of* and intuition *that* for the small sets we have been considering should then emerge in a fairly clear way. We then briefly summarize our view before going on to consider large sets and the processes involved in obtaining a hierarchy of finite sets.

Because we are to think of the intuition of a set as a single object as founded on the intuition of its elements we should pause to consider the special cases of the empty set and the one-element sets. Our remarks here can be somewhat analogous to what was said earlier about the numbers 0 and 1. In the case of unit sets there is no question about synthesizing items in a whole. We have the limiting case of the condition that the items be "synthesized", i.e., the condition imposes nothing. Nonetheless, by virtue of the process of collecting that we have discussed there are conditions in our experience for constitution of unit sets. For consider the case where we are running through a group with more than one object. Then at the stage where we take account of the first object a distinction will be constituted between the consciousness of that object and the consciousness of the other objects which would be merely protended. This distinction is enough to found the intuition of the unit set in an act of reflection, in accordance with our discussion above. The unit set is the set that is intuited when the intention that there is a set (i.e., a multitude) with at least one element is fulfilled but the intention that it contains more than one element is frustrated. In this case we still have an intention directed to a set but the set contains only one object. Thus there is a quite natural basis for the idea of intuition of unit sets in our mathematical experience.

What about the case of the empty set? We cannot of course suppose that there is an intuition of the empty set founded on the intuition of its elements because it has no elements. Hence, there is no intuition of the empty set. However, to say that the intention to the empty set is "fulfilled" is to say that the intention expressed by $(\exists x)\ (x \in X) \rightarrow \perp$ is fulfilled, i.e., $(\exists x)\ (x \in X)$ is frustrated. Proof that this is so would generally involve the construction process for sets. This just corresponds to the situation in our experience where our intention that there is at least one object in a set contains an absurdity. We no doubt sometimes suppose that there is a set of objects satisfying some condition without knowing beforehand whether there exists a single object satisfying the condition. It then later turns out that we come to know that no object satisfies the condition, e.g., the set of rational numbers x satisfying the equation $x^2 = 2$. What we should say about this example is that there are no rationals satisfying the condition, and hence there are no objects out of which we could form a set corresponding to the condition. Thus we should take the symbol "Ø" to be an expression for this particular kind of complex experience.

In our formulation of HFI we shall therefore understand "Ø" in terms of these considerations. The "pure" HF sets would be obtained by leaving

individuals out of consideration. Leaving individuals out of consideration would evidently amount to abstracting from the "origin" of the intuition of sets on a phenomenological view. Of course in applications of set theory one would generally want a formulation that provided for individuals. But for the pure HF sets the iteration of the construction process for sets begins simply with Ø. We shall have more to say about the iteration of the construction process in a moment.

We have been saying that the items we run through are contained as *parts* of an act at a given stage, so that it might appear on this conception that the relation of element to set is to be understood along the lines of the relation of part to whole. One might object to an account of (intuition of) sets which implies such a kinship on several grounds. Husserl however was already able to distinguish between different kinds of part/whole relations and certainly to recognize a difference between mereological sums and sets. This is perhaps indicated most clearly by his discussion of the constitution of sets in *EJ*.[12] It is clear from his writings, moreover, that we should regard existence assumptions about finite sets, not just mereological sums, as knowable. Husserl's interest in what we would now call mereological sums was certainly not motivated by nominalist aspirations, as has been the case with Lesniewski, Goodman, and others. In fact, Husserl argues against nominalism in mathematics, as we have noted. Just what this entails about finite sets that takes us beyond what the mereological nominalist would recognize as evidence for the existence of objects can be illustrated in a straightforward way. For suppose we consider a case where we are given a small number n of items. Then the mereologist will be prepared to recognize the existence of 2^{n-1} sums (wholes) which have these items as constituents. No more sums could possibly be obtained from these items. From the same number of items we should be prepared to recognize the existence of $2^n - n - 1$ sets, where we do not count unit sets or the empty set. But now let us suppose that the sets so constructed could themselves be viewed as distinct items to be collected into further sets. The sets obtained as a result of this process could then themselves be looked at as items to be collected into yet further sets, and so on. In supposing that we have evidence for the existence of such objects we clearly move beyond what the mereological nominalist is prepared to acknowledge. For, as we have been saying, we should be prepared to recognize that there is evidence for the existence of such objects insofar as there is for a mathematician M a well-determined process of construction for obtaining the objects. The intentions directed to such objects need not be actually fulfilled, but they should be fulfillable. Partial fulfillment of the intentions to such

objects constitutes intuition of the objects in the same sense that partial fulfillment of intentions directed to ordinary perceptual objects constitutes intuition.

We should note that thinking of finite sets as certain kinds of "wholes" composed of "parts" is not entirely inappropriate. For HF sets, and even for the view of sets in general as forming a cumulative hierarchy, we are speaking of well-founded sets. However, I think that more interesting questions about viewing sets as wholes composed of "parts" arise when we move beyond HF sets to transfinite sets, especially nondenumerable sets, as these are classically conceived. For talk of the "parts" of such sets, where we wish to view intuition of the whole as founded on intuition of the parts, becomes problematic.

It should be clear how we shall understand various set-theoretic operations and relations like $X \times Y$, X^2, X^3, $X \cup Y$, $X \cap Y$, $X - Y$, $X \subseteq Y$, $X \supseteq Y$, and so on, in the case of quite small finite sets. We view them as mental processes that are to be carried out if the intentions to the particular sets involved are to be fulfilled, or if the intention *that* a particular set stands to another set in a certain relation is to be fulfilled. The processes involved in intuition *of* a particular set are thus readily distinguishable from the processes involved in intuition *that* in the case of singular statements about sets.

In the case of the small finite sets we have been concentrating on we can therefore readily view set-theoretic knowledge as a product of (empty) intentions and intuition. M's belief in the existence of such a set is produced by a construction process which gives evidence for it. For small enough cases the construction process can actually be completed. The construction process is repeatable at different times and places and by different people. In some cases the evidence for the existence of the set would perhaps be adequate, and we might also consider it apodictic since we cannot imagine what it would be like not to be able to carry out the relevant acts (except under very pathological conditions). Similar considerations apply for intuition *that* in the case of some singular statements about some sets. We should not however suppose that we have such evidence with respect to knowing all of the possible properties of a given set or all of the relations in which it stands.

7. LARGE SETS AND A HIERARCHY OF SETS

M's belief about the existence of a large finite set is of course also produced by a construction process which gives evidence for it. But in this case M's intention can actually be only partially fulfilled and a condition for regarding

it as fulfillable would be that the process be known to be well-determined. Knowledge that the process is well-determined will involve acts of reflection which have the construction-process for finite sets as their object. For large sets the horizon of the act's intention and various background beliefs about the object play a more important role in determining the intuition of the object. The intuition of the set as a whole will be founded on intuition of some of its parts. Hence, the evidence provided by intuition for the existence of large finite sets will not be of the same character as that for quite small finite sets, but it will still bear a direct analogy to the evidence provided by straightforward intuition for the existence of ordinary perceptual objects.

Once we suppose that sets of individuals have been constructed from a given domain D of individuals, and in the case of quite large sets, that they are at least constructible, we can begin to construct a hierarchy of finite sets in a quite natural way on the basis of the processes described, so that we shall be able to regard HF(D) as constructible in intuition. Given some individuals we can collect them and become aware of them as a set. If there are a great many individuals we think of the set of them as constructible since we cannot actually complete its construction. Nonetheless the intention directed to this set can be at least partially fulfilled. We might then suppose that we construct not only this set but all possible sets of the individuals. To each of these construction-objects corresponds a distinct construction-process. There are thus collectings, and corresponding to them, collections. We want to think of the awareness of collections as awareness of abstract objects, but there is a modal element in our description of collectings, for most collectings can only "possibly" be carried out.[13]

In carrying out the process of constructing all of the possible sets we of course do not actually complete the construction of those sets which are too large. The sets "transcend" our actual experience. It will also happen that we do not wish to or cannot actually construct *all* of the possible sets of individuals. However we can actually construct at least some of the different sets of individuals and we can clearly regard the remainder as constructible. The idea of constructing all possible sets of individuals in this context poses no special problems since we know how this process would go. Now by acts of reflection we could regard the sets so constructed as themselves objects to be collected. Hence we can collect them and become aware of them as a set. Or again, we might take all possible collections of them, regarding each of these collections as a set. All of our remarks about what can actually be constructed and what is to be regarded as constructible apply again at this stage. We might again, by acts of reflection, take the sets constructed at this

stage of the process to be given objects that can be collected to construct yet further sets.

The process we are describing can obviously be continued, giving us "higher-order" constructions of sets founded on objects that are constructed, or could be constructed, at earlier stages in the process. (Note that there is no reason why we could not take this hierarchy to be a cumulative hierarchy.) Since as a rule we cannot actually complete the construction of all the sets involved at a stage in this process it is clear that what we are really concerned with as we continue through the stages is the nesting of the construction-processes in other construction-processes. The fact that we suppose there is a nesting of processes in this way, even though we do not suppose the processes will actually be completed, would of course also have applications in other parts of mathematics so that we could claim there is a foundation in intuition for more sophisticated parts of mathematics. For the idea is that it must be *possible* to carry out the steps in the intuition of intended objects. We suppose that more and more extensions could be synthesized. The *actual* intuition will not be as important but it will nonetheless provide the foundation for the entire edifice since what is to be regarded as "possible" must be founded on what we can actually carry out. That is, we must have evidence, produced by reflection on what we can actually do, that it is possible to continue, and, if we are to have an even better grade of evidence, that the way we would continue is well-determined. In the hierarchy of sets then we could regard particular construction-processes for the different sets as objects in acts of reflection. Then the construction-processes involved in forming different collections of the construction-objects can be regarded as objects in acts of reflection. They provide the representing content for the acts of reflection, and then the process continues. In each case acts of reflection take as their objects a previous process, as we have been saying. Hence an intention to a set somewhere in the hierarchy can be at least partially fulfilled, and background beliefs will function to a greater or lesser extent to fill in other indeterminacies in the knowledge of it. One might say that our knowledge here depends more on conception than on actual intuition, but it could not be regarded as knowledge at all if there were not some intuition process that could be carried out for these objects. The fulfillment of the intention obviously involves a hierarchy of founding and founded acts. For the process of obtaining the object would involve working through founding and founded acts until we arrive at the object just as it is intended. The process for obtaining the object will be well-determined.

The stages in the process are obviously going to be ordered. At the earliest

stage we construct all possible sets of individuals from a given finite domain D. At later stages we construct all possible sets of objects obtained at earlier stages. If we do not suppose that D is finite then the power set operation would have to be restricted so that it applied only to finite subsets of D. The idea of "earlier" and "later" stages here can be understood along the lines of our previous discussion of the temporal structure of consciousness. We might think of the stages as indexed by finite ordinals so that we have the familiar picture of the sequence of stages R_0, R_1, R_2, ... Where D is finite (otherwise we would need to indicate restriction on forming power sets) this is of course given in a general way by recursive definition:

$$R_0 = D$$
$$R_n + 1 = \mathscr{P}(R_n).$$

We may include Ø among the items to be collected into further sets since there is a foundation for it in our experience. For the pure HF sets we just have $R_0 = \emptyset$ and the construction is then quite simple. On the view we have developed then we should be able to regard the elements of HF(D) as constructible in intuition on the foundation of construction of (initial) parts of it in intuition. Of course we do not suppose that we actually obtain R_ω in this process.

The question of how we know that we *can* continue this process can be answered as we answered it in our earlier discussion. We have evidence that it can be continued which is based on what we can actually do, and on seeing how the process would go by reflecting on what we can actually do. The process of building up the hierarchy can itself be taken as the object of acts of reflection. It is only through acts of reflection on it that we can determine what is true or false of it, for in acts of reflection intentions which have the process as their object will be fulfilled, partially fulfilled, or frustrated. Acts of reflection will thus produce the evidence for beliefs about the process.

It seems to me that our description of the construction of the HF sets illustrates clearly in a particular context the Husserlian theme, taken over from Kant, that knowledge begins with sense-experience but is not limited only to sense experience. For on the view we have developed set-theoretic knowledge has its origins in acts of straightforward perception directed to individuals. That perceptual individuals can be said be immediately given in ordinary sense-experience is of course a fundamental phenomenological thesis. When properly understood it is a thesis that would be difficult to challenge. Objects like sets however are not given in straightforward perception. Properties of perceptual objects are quite different from

properties of sets. Sets are not located in space-time, cannot be moved about, do not have color, and so on. But knowledge of numbers and finite sets is founded on the process of perception by acts of reflection and abstraction. Our knowledge is not limited to sense-experience because of acts of reflection and abstraction on the processes involved in sense-experience. There are however still constraints on what can be regarded as knowledge because what is obtained by reflection and abstraction must be founded on what has already been given. There is, so to speak, a closure of knowledge under the processes recognized. The knowledge founded on given experiences by certain admissable processes is built up inductively so that anything not obtained by the processes cannot be presumed to give us knowledge. In the context of set theory it is possible to get an especially clear view of the Husserlian claim that the processes which give evidence for M's belief in the existence of mathematical objects must be founded on processes that are previously given, and ultimately on the process of perception. For by disregarding the claim we can easily be led into error or illusion about what kinds of sets exist. On our conception certain sets could never be formed, e.g., a set which contains all sets, a set which contains itself. Disregarding the claim would be tantamount to supposing that we can *know* that certain "objects" exist even when they have not been obtained by construction from previously given objects. The idea that processes can be applied only to what has already been given has of course been a cornerstone of constructivist conceptions of mathematics for some time. Our description of the construction of HF sets illustrates the older Kantian theme in a way that shows its close relationship to constructivism in mathematics.

Of course on our interpretation what gives substance to this theme is the recognition of the structure of founding and founded acts, and in particular of acts of reflection and abstraction. Husserl claims that all of our higher, theoretical forms of knowledge are based on just such hierarchies of acts. As we mentioned earlier, Husserl believed that he had improved upon Kant in recognizing the role of founded acts in providing knowledge, and in recognizing the corresponding extension of the conception of intuitive processes that must go with this. Short of recognizing these processes as *intuitive* and investigating them it is not clear how Kant had grounds for supposing that *knowledge* was not limited to just what could be given in sense-experience (i.e., in straightforward perception). Husserl wished to have an account of knowledge that brought together the view of knowledge as a product of (empty) intentions and intuition and the view of knowledge as beginning but not ending with sense-experience.

8. ILLUSION IN SET THEORY

Since we mentioned questions about illusion or error in set theory we might recall the analogy Gödel draws between set-theoretic paradoxes and deceptions of the senses.[14] What Gödel suggests is that in the case of mathematics our conceptions about objects are changed and adjusted through further experience, just as they are in ordinary perception. The only way we can hope to dispel illusion is through further experience. One might argue that Russell's paradox provides an extreme but clear case of this in mathematics. For in the case of Russell's paradox the assumption that a particular intention can be fulfilled, or that there can be evidence for an object (an extension or set) corresponding to a particular predicate, leads to absurdity. The intention to "the set which contains all and only those sets that do not contain themselves" cannot be even partially fulfilled. The reasoning that shows this can be carried out in very restricted systems of logic, e.g., the reasoning does not require use of the law of the excluded middle.[15] Then where we have a naive comprehension principle of the form

$$(\exists y)\ (\forall x)\ (x \in y \leftrightarrow P(x))$$

it is clear that there could not be a set, or an extension of a predicate, corresponding to the predicate $x \notin x$. For if so we would have for some a that $(\forall x)\ (x \in a \leftrightarrow x \notin x)$, and so in particular $a \in a \leftrightarrow a \notin a$. After it was realized that the assumption that the intention to this object could be fulfilled led to absurdity adjustments were made in the conception of set which gave rise to the absurdity. One way to look at this would be to say that it could no longer be supposed that every predicate has an extension, or that to every predicate there corresponds a set. The intentions expressed by some predicates may be objectless in principle. We might say that the discovery of the illusion of thinking that there could be a universe of sets among which would be included the Russell set or other troublesome sets led to a change in what was meant by the term "set", i.e., to a change in the earlier intention. Beliefs about sets change just as beliefs about perceptual objects change when intentions involving the objects are frustrated. The intention is corrected or adjusted in light of further experience. In both mathematical and perceptual experience there is an interplay between conception and intuition in the acquisition of knowledge.

9. CONCLUDING REMARKS

To conclude our brief reflections on finite sets we should recall several remarks made earlier about the intuition of natural numbers that will have application here. First, it could be argued that we have grounds for speaking of finite sets being given as abstract objects due to the structure of founding and founded acts. But this should be understood in terms of our remarks about the phenomenological reduction: we are concerned with an analysis of the acts and processes in which there is constitution of the consciousness of the existence of finite sets, relations between finite sets, and so on. Following Husserl we could then think of the "meaning of the being" of finite sets as derived from these acts and processes. Carrying out the program of analyzing these processes in mathematics generally may be a vast undertaking but it could evidently be undertaken independently of traditional metaphysical questions about the existence of abstract objects so that we need not see the kind of "idealism" entailed by the phenomenological reduction as necessarily incompatible with metaphysical realism about the objects. We can also respond to problems about the "completeness" of finite sets and related problems posed by a structuralist understanding of mathematical objects. For we do not identify finite sets with any particular representations of them but instead point out how on our conception set-theoretic intuition has the function of individuating sets by virtue of the processes in which they are constructed. We can carry out acts of intuition that enable us to identify sets in relation to one another. Particular sets can be viewed as objects in terms of the places they occupy in the structure that is generated. We might also recall our earlier comments about mathematical truth "within its horizons". We shall have more to say about this and about some of the other matters just mentioned in Chapter 8.

Let us close with a few remarks on the role of intuition in providing evidence for the truth of *general* statements about HF sets. Our remarks here could be developed along the lines of our remarks concerning general statements about natural numbers. In particular, proofs using the induction axiom will parallel the construction process for finite sets. Hence they depend on the insight that it is possible to continue the construction process. This insight could only be produced by acts of reflection which have the construction process as their object. It is in acts of reflection that we would have verification of the fact that it is possible to continue the construction. But acts of reflection only produce such verification by being fulfilled or partially fulfilled, i.e., by intuition. Hence, the evidence for the truth of general

statements proved by induction will be founded on intuition. We could make remarks about the evidence provided by intuition for general statements on the analogy of our earlier remarks about number theory. We note in particular that we do not assume that intuition gives knowledge of truth in an absolute sense.

It should now be possible to better understand how mathematical intuition provides evidence for the truth of the axioms of finite set theory that we gave above. By the same token we could consider the usual axioms of ZF minus the axiom of infinity.

CHAPTER 8

CRITICAL REFLECTIONS AND CONCLUSION

1. INTRODUCTION

In this chapter we shall, by way of conclusion, briefly take stock of how the account of mathematical intuition we have been developing avoids a number of the usual objections to the notion of mathematical intuition. Since much of what we have said about mathematical intuition provides only an outline we note where the account should be further developed to avoid other possible objections. We shall also briefly comment on some of the larger issues that lie in the background of our discussion. In particular, the phenomenological view we have been developing suggests a certain perspective on some issues about platonism and constructivism, and on the question of whether it is possible to have an account of mathematical knowledge that does not preclude an account of mathematical truth. We should also make some comments about how this approach is to be understood in connection with the parts of mathematics we have not considered.

2. SUMMARY OF THE ACCOUNT

On the view we have been developing the function of mathematical intuition is to provide evidence for M's belief that S, where S is a statement expressing some mathematical intention – in particular, an intention involving natural numbers or finite sets. Mathematical intuition provides evidence for the truth of singular statements about such objects, and also for the truth of general statements about such objects, by virtue of acts of reflection and abstraction. Intuition *that* some state of affairs holds for natural numbers or finite sets goes hand-in-hand with intuition *of* these objects. Intuition in the case of particular mathematical objects has a *de re* character. The notion of mathematical intuition is thus to be understood as having a function in mathematical knowledge analogous to the function of ordinary perceptual intuition in perceptual knowledge. For in ordinary perceptual knowledge we also think of M's beliefs as produced by intuitive processes that give evidence for them. The intuitive processes in the case of perception are to be understood in terms of sequences of essentially indexical acts directed toward some object or

objects and carried out in order in time. The same processes are applied to different objects. We view the processes in much the same way in mathematical intuition. In both cases the objects generally transcend the actual intuition. The intuition of an object as a whole is founded on the intuition of its parts. The horizon of the acts involved, determined by the relevant background beliefs about the given objects, serves to make the intuition more determinate. There is an analogy involving the "adequacy" of the evidence given by intuition in each case. Note that in either case we would say that M does not know that S when M's belief that S is not produced by a process that gives evidence for it, i.e., when intuition is not involved, since intuition is what provides evidence. If M could not have even a partial intuition we would not say that M knows that S. For then M's belief that S would be held independently of being able to undertake some process in which evidence for it could be accumulated. We would regard it as "mere conception".

The kind of analogy we are suggesting between mathematical and perceptual intuition should figure more generally into an account of knowledge in both the mathematical and physical sciences. We have said little about the physical sciences but have usually spoken instead of "ordinary" perceptual intuition, or of "medium-sized" physical objects. But surely we can also view the statements of physical theories as expressions of intentions. The question then arises of whether and how these intentions are fulfilled. The fulfillment will again need to be appropriate to the intention. Different types of evidence will correspond to different types of objects. This will be especially important to note in the case of physical theories that speak about objects that obviously cannot be perceived by ordinary perceptual means, eg., the "objects" of microphysics, the core of the sun, the "objects" of cosmological theories. For the objects posited by such theories are of a different type from ordinary perceptual objects. They are postulated, for example, as objects existing in very small regions of space-time, or at very great distances from us, as objects that are not available in ordinary sense perception. Given our powers of discernment and other structural limitations on perceptual ability it seems impossible that we should perceive such objects by means of ordinary sense perception. It could still be argued that there is a distinction between mere hypothesis and observation for these objects, between knowledge and mere conjecture, and on this score our observations about analogousness will still hold. For even though the objects are of a different type from ordinary perceptual objects, just as mathematical objects are also of a different type from ordnary perceptual objects, there is a sense of "observation" appropriate to such objects. "Observation" here will still

consist of *processes* for verification of hypotheses. We develop procedures or processes appropriate to the observation of such objects. These processes are built up genetically over time as our intentions regarding such objects are built up, refined, and reshaped in the growth of knowledge. The processes will provide evidence that is more or less adequate, that is intersubjectively shareable, and so on. Husserl would no doubt say that this kind of observation is "founded" observation, invoking his conception of the Lebenswelt as the world of everyday perception which is presupposed even by highly sophisticated forms of scientific observation. It is presupposed at least in the simple sense that without ordinary perception we would not have observation in the physical sciences either. What is important for the analogy from our point of view is the fact that such processes exist in the physical sciences, and that we continue to develop and refine them. This clearly bears an analogy to our view of intuition in mathematics.

We can therefore claim that there is an analogy between perceptual and mathematical intuition insofar as in each case we are concerned with certain *processes* that produce evidence for M's beliefs in these respective domains of knowledge. The objection that mathematical intuition could not be a source of knowledge about objects, analogous to perceptual intuition, can be set aside. For the objection depends on the observation that mathematical objects like natural numbers and finite sets are abstract objects which cannot be ostended, cannot be located in space-time, cannot be causally related to our senses, and are not individually identifiable and determinate, as are concrete perceptual objects. But on our view the analogy between mathematical and perceptual intuition does not concern the objects involved in either case. Rather, it concerns the structure of the acts and the processes in which the objects are given. We have in fact continually noted that mathematical and perceptual objects are objects of different categories, each category having its own type of evidence. Mathematical intuition is of genuinely mathematical objects, not of perceptual objects like sign tokens, that might serve to represent natural numbers or finite sets. Perception of the latter kinds of objects could never yield knowledge of mathematical properties, except on pain of circularity. What is needed to give substance to the view that mathematical intuition is of mathematical objects like natural numbers and finite sets, not of sign tokens or other physical representations, is the idea that acts of mathematical intuition, unlike straightforward perceptual acts, are founded acts. What this means is that they involve, in the first instance, a change in attitude (thetic character of acts directed) toward particular objects that may be given. What changes in seeing a particular number of perceptual

objects *as a set* is our awareness, the way we regard the individuals. For we are then no longer seeing the individuals as straightforward perceptual objects but rather as elements of a set, where the particular nature of the individuals is irrelevant to the set. This shift of awareness is due to founded acts of reflection and abstraction. The awareness of mathematical objects, being a founded awareness, is not "immediate" but rather must be constructed. The objections mentioned earlier that are sometimes raised to the notion of founding can thus be avoided on our view.

We have also tried to show how knowledge about numbers and finite sets obtained by mathematical induction is founded on intuition. Here it is pointed out that what needs to be seen is that a relevant construction process exists and *can* be carried out, for proofs by induction will parallel the construction process. Insight about how the construction process behaves could only result from reflection on the construction process, from acts which have the construction process as their object. On the basis of these kinds of acts we could claim that we have evidence that general statements proved by induction are fulfillable. We do not suppose that acts of reflection on the construction process, being founded on what we can actually carry out, provide truth in an absolute sense.

3. AREAS FOR FURTHER WORK

We shall not review here any other objections and responses that have been mentioned in the course of our discussion. Instead we might simply note that more should be said about several philosophical matters important to our view. Perhaps most importantly, a fuller account of acts of categorial abstraction and of acts of reflection should be pursued. We have also not gone into questions about degrees and types of evidence in very much detail but to do so would be useful for making distinctions in the kind of knowledge a mathematician M can be presumed to have in mathematics. We have tended to concentrate on general conditions involved in the distinction between empty intentions and fulfilled or fulfillable intentions without considering finer details. Other matters that could be gone into in more detail concern the relation of intuition *of* to intuition *that*, and the distinction between *de re* and *de dicto* intuition; further questions about intuition and structuralism; the comparison with Kant's views; and questions about intuition and transfinite objects. Investigation of questions in the latter area might help to illuminate and to set in clearer relief what has been said in this essay.

We should also say something about how our view is to be understood in

connection with parts of mathematics we have not discussed. We have deliberately focused on very elementary parts of mathematics with the idea of exploring in this context some of the central features of a phenomenological conception of intuition. The approach to other areas of mathematics would generally take the form of the strategy we have followed. That is, start with the observation that real analysis, complex analysis, functional analysis, geometry, topology, abstract algebra, category theory, and so on, exist. Insofar as we are willing to suppose that results in these areas provide some degree of knowledge we can ask how that knowledge is possible. This would involve an analysis of "origins" in the sense in which we have used the term throughout this essay. If we are to regard ourselves as having knowledge of objects in these areas then intuition is a necessary condition for that knowledge. Mathematical statements from any area of mathematics can be viewed as expressions of intentions, and as such they may be fulfilled, partially fulfilled, presently unfulfilled or even provably empty. If an intention is to be regarded as fulfilled, or as fulfillable, there must be a construction (intuition) process that one could undertake to find the object(s) toward which the intention is directed. This admits some degree of latitude in evidence, but it would still generally tend to attach greater value than has heretofore been attached to constructively acceptable results in these areas of mathematics. Thus our approach would extend to mathematics as a whole in much the same way that Bishop and other constructivists have attempted to treat different areas of mathematics constructively. In Bishop's work, for example, there is a constructive development of calculus and the real numbers, set theory, metric spaces, complex analysis, measure theory, locally compact Abelian groups, and commutative Banach algebras. There is of course some question about just which kind of constructive mathematics we are speaking about. We have not had to seriously address this question because we have been dealing with elementary parts of mathematics on which most constructivists (save ultrafinitists) would agree. There are many questions about the development of higher parts of mathematics that could be explored here: for example, should we follow Bishop's practice, should we adopt Markov's principle, what is the status of the "creative subject", Kripke's schema, bar induction, choice sequences?

It might be argued that the answers to these questions will also determine, at least in part, the attitude that should be taken toward classical mathematics. However, I do not think it follows from anything we have said thus far that classical mathematics is to be rejected. I do not see the premisses that would lead to such a conclusion coming from phenomenology.

4. PLATONISM, CONSTRUCTIVISM, AND BENACERRAF'S DILEMMA

In conclusion, let us consider in more detail some remarks that have been made about the meaning of the phenomenological reduction in mathematics, especially as it applies to questions about idealism and realism. We shall then briefly remark on some consequences of the idea of mathematical truth "within its horizons".

What has been said about natural numbers and finite sets as "objects" of intuition should be compared to our remarks on physical objects and the phenomenological reduction or epoché. We said that from the point of view of phenomenology we do not need to consider anything more than the structure of the acts and sequences of acts in which objects are given, whether the objects be physical or mathematical. We are not to suppose that there is some kind of relation independent of these processes which puts us in touch with objects. What phenomenology attempts to do is to analyze the experience or act-structure which is necessary for knowledge of objects. This kind of undertaking might suggest that phenomenology is committed to a version of idealism regarding mathematical objects. As Aron Gurwitsch has put it in a discussion of Husserl's views on logic and mathematics,

> Philosophical reflections on logic... have in fact led Husserl to establish the principle of *phenomenological idealism* according to which whatever exists and has validity derives the sense of its existence and its validity from conscious life and can find its ultimate clarification and final justification only by means of analyses of acts and groups of acts in which it presents itself as existing and as valid.[1]

For mathematics as a whole phenomenology is thus thought to suggest a vast research program into the acts and processes involved in the constitution of different kinds of mathematical objects and the truths about them. It would not be too far off the mark to say that mathematical investigations of the relevant kind have been carried out in constructive mathematics. As we mentioned earlier, however, we might raise questions about whether the account we have given of the intuition of natural numbers and finite sets is compatible with the view that these are mind-independent objects about which statements might be true or false independently of our knowledge. For if so, the view we have been developing might lead to a more refined understanding of the relationship of constructivism to platonism in mathematics.

It could be argued, I think, that the idea that natural numbers and finite sets

are mind-independent objects is compatible with a phenomenological analysis of the intuition of such objects. We have been saying that mathematical intentions can be viewed as referring to objects thought to exist independently of us but also to expectations about the course of our experience regarding such objects. Depending on which way we view mathematical intentions we could expect different research programs to develop in mathematics, and there seems to be no compelling reason to legislate in favor of one attitude at the expense of the other. In the one case the concern is with M's mathematical knowledge, with the fulfillability of the intentions, while in the other case questions about the fulfillability of the intentions are not relevant. Both attitudes are found in mathematical practice. In considering mathematical intentions in terms of the course of our experience with respect to natural numbers and finite sets we are attempting to analyze sequences of acts by which we obtain knowledge of objects, independently of naive metaphysical questions about the objects. To investigate constructions does not require that we take a stand one way or the other on the question of the existence of numbers and finite sets independently of the way they appear in consciousness. For as we have depicted the phenomenological reduction it is not claimed that the existence of a number or finite set depends on the possibility of undertaking a construction. Being able to carry out a construction explains our ability to have knowledge of such objects.

It should be noted that we are not saying, as is sometimes said in constructivist developments, that numbers and finite sets are our own mental constructions, so that they could not be taken to exist independently of these constructions. On the contrary, the constructions themselves are mental processes, but numbers and finite sets are not given as our own mental constructions. It would amount to a kind of gross psychologism to impute to numbers and finite sets properties of mental objects. Constructions are given as having temporal duration, for example, but numbers and finite sets are not. As Husserl puts it in *EJ*, intuitions are temporal processes but objects like natural numbers and finite sets that are given in these acts are given as "omnitemporal".[2]

Husserl offers arguments in a number of places in his later writings to show that mathematical objects could be taken to "exist" independently of our minds, or of our constructions. In *EJ* we find him saying, for example, that

> 'there are' mathematical and other irreal objects which no one has yet constructed. Their existence, to be sure, is revealed only by their construc-

> tion (their 'experience'), but the construction of those already known opens in advance a horizon of objects capable of being further discovered, although still unknown. As long as they are not discovered (by anyone), they are not actually in spatiotemporality; and as long as it is possible (how far this is possible, there is no need to decide here) that they will never be discovered, it may be that they will have no world-reality.[3]

In other passages we find Husserl saying that in any case mathematical objects are given as independent of us in the very acts in which we come to know about them. They acquire this "meaning of being" in our acts. On this kind of view constructivism is therefore not necessarily incompatible with the idea that there are mathematical objects that exist independently of our minds, and that statements about them may be true or false independently of our knowledge.

Much more should no doubt be said about this kind of view. It would seem to be consistent, for example, with a number of remarks that Gödel makes about platonism in "What is Cantor's Continuum Problem?" and "Russell's Mathematical Logic". It does not seem to be readily assimilable to traditional intuitionism as represented by Brouwer and Heyting. Both Brouwer and Heyting speak somewhat uncritically of mathematical objects as mental objects. It seems to me that a more fully developed phenomenological view would entail a critique or a correction of intuitionist *philosophy* on several counts: (i) on the understanding of constructions in the sense of "objects" obtained as results of construction processes, and in the sense of construction-processes as "objects", (ii) on questions of idealism, psychologism, and solipsism in mathematics, and (iii) on the theory of meaning of mathematical statements.

A further consequence of the phenomenological reduction of which we should take note is that we do not suppose in the case of either mathematical or perceptual intuition that there is any kind of causal relation between the object and the sensory mechanisms of the perceiver. The point of the phenomenological reduction is just that we are to investigate the structure of acts and sequences of acts, and to do so does not require considerations about causality. Since we are not committed to providing an analysis of the condition that the state of affairs referred to by S *causes* M to believe that S we can avoid a difficulty which plagues many attempts to develop the idea of mathematical intuition: The difficulty that there could not be anything like mathematical intuition, analogous to perceptual intuition, because in perceptual intuition we are causally related to objects, while in mathematical

intuition this would be impossible. On our view there is an analogy between mathematical and perceptual intuition and an analysis of any condition about causes is not involved in either case. It should be noted that we are not saying that ordinary perceptual objects are not given as objects to which we are causally related, or as objects that may be causally related to one another. On the contrary, they would certainly be recognized as so given, unlike mathematical objects. The idea is rather that in the case of either mathematical or perceptual intuition it is the structure of the mental acts and processes in which objects are given that is to be investigated. This approach is especially important for giving an analysis of mathematical knowledge because we do not think of typical mathematical objects as objects to which we could be causally related. Instead of attempting from the outset to apply a causal account of knowledge where it does not seem to fit – in mathematics – we concentrate on the mental processes that M can undertake in mathematics and that provide evidence for M's belief that S where S is some mathematical statement. We have described in particular the processes involved in providing evidence for the existence of natural numbers and finite sets. It is evident from what we have said that the investigation of *constructions* does not require any kind of causal analysis. Indeed, a significant amount of investigation of mathematical constructions has been carried out in mathematical practice by intuitionists and other constructivists. It is clear that constructions can be studied quite independently of causal considerations and that, properly interpreted, they can be thought of as the processes that provide mathematical knowledge.

On the view we have discussed, then, the fact that mathematical objects are thought of as "abstract" does not prevent us from having knowledge of such objects on account of lack of causal relations. In the case of natural numbers and finite sets we could fix our position in relation to Benacerraf's dilemma in the following way. We might think of natural numbers and finite sets as abstract in the way that a Tarskian account of truth would depict them. But it does not follow that we cannot have knowledge of these objects, for we do have constructions in which such objects are given. In the case of large numbers and finite sets the objects are indeed only partially given. But the important aspect of the phenomenological view is that partial intuition of objects still yields knowledge of the objects, just as in perception. The conclusion to be drawn is that partial intuition yields knowledge, but that the knowledge it yields is "inadequate". Mathematical intuition provides different grades of evidence for the truth about abstract objects like natural numbers or finite sets, it yields better or worse approximations to the truth

about such objects. We do not identify truth about such objects with intuitability or, e.g., assertability. This gives a different perspective on the Benacerraf puzzle, for we do not suppose that intuition can give maximal or complete knowledge of arithmetical truth or set-theoretic truth. We should recall in this context Husserl's view of truth as a regulative idea (in a Kantian sense) "lying at infinity". We might think of a Tarskian account of truth as giving what from the point of view of epistemology we might call "ideal" or "absolute" truth conditions – truth conditions which correspond to ideal or maximal knowledge. It represents truth in conditions under which our knowledge would be complete. As we mentioned earlier, this amounts to conditions under which every mathematical statement could be decided in intuition. We might think of the Tarskian theory of truth as abstracting from actual and in some areas of mathematics even possible conditions of cognition. Following Husserl's views in *FTL* we would have to regard it as providing a concept of truth "torn out of the nexus of the living, ongoing intentional processes which aim at the perfection of knowledge". Our knowledge can only approximate the ideal to a greater or lesser degree. To expect any more in light of the limits of our intuitive abilities would be unreasonable.

Husserl discusses this last point in some detail in *FTL* when he criticizes what he calls "absolutist" theories of truth. He is very concerned to point out that there is no absolute evidence which would conform to a notion of absolute truth. Any truth we can experience will be involved in relativities by its very nature. It is in this context however that he asks the question we mentioned much earlier in this essay:

> What if the relativity of truth and of evidence of truth on the one hand, and on the other hand, the infinitely distant, ideal, absolute truth beyond all relativity – what if each of these has its legitimacy and each demands the other?[4]

In the case of mathematics the implied answer to this question suggests a general way of viewing the relationship of a platonistic conception of mathematical *truth* to a constructivist conception of mathematical *knowledge*: a conception of ideal or maximal truth serves as a regulative idea, as an ideal or norm toward which knowledge aims if it is to be more fully determined and perfected.[5] One might look at a platonistic view of truth as *postulating* truth as fully determinate, complete, perfected. Kant would say that "reason" (as distinct from understanding) inevitably leads to such postulation, and that it thereby provides a standard against which our knowledge is measured at

any given time. But at any given time we understand far less than we can think or reason about. Even the most highly developed sciences, Husserl argues, are involved in relativities by their very nature, and are referable to regulative ideas as their norms. In his critique of logical and mathematical knowledge and more generally, of the "exact" sciences, Husserl thus adopts the view according to which

> we have truth not as falsely absolutized, but rather in each case, as within its *horizons* – which do not remain overlooked or veiled from sight, but are systematically explicated. We have it, that is to say, in a living intentionality (called "evidence of it") whose own content... with the uncovering of the attendant intentional implications, leads to all those relativities in which being and validity are involved.[5]

Viewing mathematical truth as "truth within its horizons" permits us to avoid "skeptical relativism" about mathematical knowledge on the one hand, and any kind of logical or mathematical "absolutism" on the other. What we have been saying reflects this position directly. For the position we have been discussing is far form being an epistemologically absolutist position. We do not suppose that even elementary arithmetical or set-theoretic knowledge is absolutely certain and secure. We simply do not and cannot have the evidence to guarantee that. On the other hand, we *do* have evidence for the truth of various arithmetic and finite set-theoretic statements, on the analogy with perception. It is not as if we must adopt the view of the skeptic or relativist about mathematical knowledge. Neither skeptical relativism nor absolutism could hope to give us an accurate picture of mathematical knowledge.

NOTES

CHAPTER 1

[1] For a recent discussion of geometric intuition from a phenomenological point of view see Tragesser, R., *Husserl and Realism in Logic and Mathematics* [137]. Also, [136]. Also Husserl's unpublished studies [57], [69], and Becker's study [3].
[2] In a Spring 1986 talk at Columbia University and in conversation at that time Dagfinn Føllesdal made a number of important points about how phenomenology should not be viewed as a foundationalist enterprise. In earlier writings and in this essay I have taken much the same view.
[3] See Steiner, M., *Mathematical Knowledge* [129], pp. 117–118, 130–132, and Parsons, C., "Mathematical Intuition" [105], p.146.
[4] See Parsons, Steiner, ob. cit. On *de re/de dicto* intentions in Husserl see Smith, D. and McIntyre, R., *Husserl and Intentionality* [128].
[5] Important interpretations and developments of Kant's views are to be found in Parsons' *Mathematics in Philosophy* [107], his "Arithmetic and the Categories" [101], and in papers by Jaakko Hintikka and Carl Posy.
[6] In *From Frege to Gödel* [143], "On the Infinite". See also "The Foundations of Mathematics" by Hilbert, in [143].
[7] "Mathematical Intuition" [105].
[8] *The Nature of Mathematical Knowledge* [77], pp. 50–53. Also "Hilbert's Epistemology" [75], and "Kant and the Foundations of Mathematics" [76].
[9] The conception of natural numbers as generalized types was first presented in "Ontology and Mathematics" [110] and is discussed further in unpublished work in progress [108]. The material on finite sets is also developed in the unpublished work. See also "Intuition in Constructive Mathematics" [103], "Objects and Logic" [109], and "Mathematics, Foundations of" [106], in the *Encyclopedia of Philosophy*.
[10] See "What is the Iterative Conception of Set?" in [107].
[11] "Kant's Philosophy of Arithmetic" in [107], p.141.
[12] See Introduction in *Mathematics in Philosophy* [107].
[13] Brouwer, L.E.J., *Collected Works, Vol. I,* [14], with slightly different versions in different papers.
[14] "On an Extension of Finitary Mathematics Which Has Not Yet Been Used" [37]. Also, Bernays, "On Platonism in Mathematics" [11].
[15] In Constance Reid, *Hilbert* [115].
[16] In "Mathematical Logic" in Saaty [81]. See also other writings by Kreisel on the topic,[82],[83], [84], and Kreisel and Feferman on progressions of theories [25], [26], [28].
[17] This is reported by Wang in *From Mathematics to Philosophy* [145]. It is by now well-known that Gödel began studying Husserl's philosophy in about 1959 and that

his interest in Husserl continued for a long time. Through discussions with Professor Wang and visits to the Gödel *Nachlass* in the Firestone Library of Princeton University I have been able to obtain a somewhat better understanding of the nature of Gödel's interest in Husserl, but unfortunately it is still very far from being complete. All of Gödel's notes on Husserl are in Gabelsberger shorthand and to date very little of this material has been transcribed. One can discern references to specific texts and passages but the comments which accompany the references would need to be transcribed. One interesting fact is that Gödel had studied *FTL* and *PITC* in more detail than I had expected.

18 In "What is Cantor's Continuum Problem?" [39]. Following passage also.

19 "Mathematical Truth" [6].

20 See her "Perception and Mathematical Intuition" [88].

21 In "Ontology and Mathematical Truth" [70].

22 In *Ontological Economy: Substitutional Quantification and Mathematics* [44].

CHAPTER 2

1 I should say at the outset that I do not wish to prejudice issues about the metaphysics of intentionality. There are many possible positions that one could take on this in the philosophy of mind. Thus, for example, speaking of intentionality is perfectly compatible with being a physicalist, a physicalistic functionalist, or an instrumentalist. For then one might suppose that we are simply adopting an "intentional stance" toward the explanation of mathematical knowledge. The language of intentionality could be viewed as an upper-level language which has an explanatory significance not found in lower-level languages, or languages at the level of "hardware". I believe that what I shall have to say about intentionality will be neutral between various metaphysical positions about cognition.

2 See especially *LI*, I, V. Also *Ideas*, *FTL*, *CM*.

3 Several studies that might be helpful for understanding my presentation are Smith and McIntyre, *Husserl and Intentionality* [128]; Miller, *Husserl and Temporal Awareness* [94]; Føllesdal's papers [29], [30], [31]; Mohanty, *Concept of Intentionality* [96], and [97], [98], [99].

4 *Ideas*, section 24. Boyce-Gibson's translation, which I modify at certain points.

5 *LI*, VI, section 45.

6 *LI*, VI, section 45.

7 *LI*, VI, section 46.

8 *LI*, VI, section 46.

9 *LI*, VI, section 47.

10 See especially *CM*, section 38. Also *EJ*. Also *Analysen zur Passiven Synthesis* [53].

11 *LI*, VI, section 57.

12 *LI*, VI, section 57.

13 *LI*, VI, section 60.

14 See e.g., Parsons "Mathematical Intuition" [105], p. 159. Also, [107], especially p. 21, 184–185.

15 *LI*, II, section 8.

16 *FTL*, section 58.

[17] *FTL*, section 60.
[18] *FTL*, section 105. Following two passages in the text are also section 105. See also sections 106–107. This late view of Husserl about truth is reminiscent of the famous Peircean view.
[19] For a discussion of these problems see Levin, *Reason and Evidence in Husserl's Phenomenology* [86].
[20] *CM*, section 6.
[21] *CM*, section 6.
[22] See, e.g., *EJ*, section 8. We have said here that the sequence of intuitions must be *completely* determined in advance. However, it seems we could also allow that an apriori element is involved in our knowledge when the sequence of intuitions is only partially fixed. This might suggest a distinction in grades of the a priori.
[23] *FTL*, sections 94–96.
[24] See Tragesser, [137].
[25] *LI*, VI, section 66.
[26] *FTL*, section 100.
[27] See especially *LI*, I.
[28] *LI*, I.
[29] *LI*, I, sections 14–15.

CHAPTER 3

[1] *Ideas*, section 44. Boyce-Gibson's and my translations.
[2] *EJ*, section 8. The following passage in the text is also section 8.
[3] *EJ*, section 8. On internal/external horizon discussed below see, for example, *EJ* section 8.
[4] See especially Parsons' "Infinity and Kant's Conception of the the 'Possibility of Experience'", [102]. Questions about continuous phenomena might arise in another way from reflections on the structure of experienced time. It would be natural to suppose that experienced time is linear and dense, but other philosophical positions could be taken. It might be claimed that the structure of experienced time is really best characterized in terms of a discrete ordering, or a continuous ordering, instead of a dense ordering. I shall not be concerned with these questions since I am only interested in discrete acts directed to individuals that are given as being in time.
[5] For a nice description of the temporal structure of straightforward perception see I. Miller's *Husserl, Perception, and Temporal Awareness* [94].
[6] *Ideas*, section 44. The following passage is also from section 44.
[7] *CM*, section 22. Also, see *Ideas*, section 142.
[8] *Ideas*, section 142.
[9] Robin Gandy, "Church's Thesis and Principles for Mechanisms" [36]; H. Dreyfus "What Computers Can't Do" [22].
[10] See e.g., *Ideas*, section 143.
[11] Parsons' "What is the Iterative Conception of Set?" [112].
[12] Parsons' "Quine on the Philosophy of Mathematics" [111], p. 191.
[13] *Ideas*, section 138.
[14] *Ideas*, section 131. See also section 129.

[15] See Kaplan [73]. Also Smith and McIntyre, *Husserl and Intentionality* [128], pp. 213–222, 290–291, 359–360, 362–369. Also, Smith [125], [126], [127]. Also, I. Miller [94] on the behavior of the determinable "x", pp. 60–80.
[16] *LI*, I, section 26. For more material on "occassional expressions" see *LI*, VI, section 5.
[17] *EJ*, section 21. A large question one could ask is just what Husserl thinks is accomplished by the description and analysis of phenomena in terms of their "origins". This question, insofar as it is significant for the main issues of the essay, will be answered by my discussion of the role of founding and founded acts in mathematical thinking. I say more about the analysis of "origins" in Chapter 5.
[18] *CM*, section 14.
[19] There are many essays on this question. Some of the best known views are discussed critically in K. Ameriks' "Husserl's Realism" [1], where Ameriks argues that Husserl was a realist.

CHAPTER 4

[1] For a discussion of these objections see Parsons, "Mathematical Intuition" [105]. Also, his unpublished work [108]. Also, Benacerraf, "Mathematical Truth" [6].
[2] These remarks might be compared with Gödel's views mentioned in Chapter 1.
[3] See "Mathematical Intuition" [105], footnote 8.
[4] See John Nolt, "Mathematical Intuition" [100].
[5] See Nolt's discussion in the essay just cited. Nolt offers a view of mathematical intuition of natural numbers that is much like the "numbers as generalized types" conception that we have been discussing, except he is prepared to say, unlike Parsons, that ostension of numbers is direct, not deferred: "When we point to a numeral...we are pointing to a number". His view seems objectionable and certainly subject to the same limitations as that view. There would be even more serious difficulties for higher forms of number and for set theory (including ZF!) where Nolt wants to treat intuition of the objects involved as based on perception and imagination of sign tokens for the objects.
[6] Benacerraf, "What Numbers Could Not Be" [7]. Benacerraf's view has been called "eliminative structuralism" by Parsons: it is structuralism without mathematical objects. Versions of non-eliminative structuralism that are closer to our view can be found in Michael Resnik's [117] and [118], and Stewart Shapiro's [122]. Other relevant comments on structuralism can be found in Parsons, "Frege's Theory of Number" in [107], "Ontology and Mathematics" [110], and "Quine on the Philosophy of Mathematics" [111].
[7] For Putnam's remark see "Mathematics without Foundations" [113]. Although I do not address the question directly in this essay it seems to me that much of what I have to say about the intuition of natural numbers and finite sets would be compatible with the "modal", or modal structuralist, picture of mathematics discussed by Putnam and developed by others. However, the motivation for developing this program on our view would not be to dispense with abstract objects, but rather it would be purely epistemological. On the "modal" picture we could think of quantifiers for natural numbers and finite sets as interpreted modally in such a way that statements about the

existence of such objects satisfying some condition are viewed as statements about the possible existence of objects satisfying purely structural conditions. Parsons has argued that the "object" picture of mathematics and the "modal" picture of mathematics could not be viewed as giving "equivalent descriptions" of the realm of mathematical facts for higher set theory. (See "Quine on the Philosophy of Mathematics" [111].) The remarks on quantifiers here should be compared with my remarks on quantifiers in section 5 where the notion of "fulfillability" of intentions to natural numbers and finite sets plays the major role.

[8] As we noted earlier, a version of structuralism somewhat like this has been defended by Resnik in [117] and [118], and by Shapiro in [122].

[9] "What Numbers Could Not Be" [7].

[10] See, e.g., Kitcher in *The Nature of Mathematical Knowledge* [77], Chapter 3, and "Hilbert's Epistemology" [75]. The objections that follow are discussed by Kitcher. Tait raises somewhat similar objections in his "Finitism" [133].

[11] Becker, "Mathematische Existenz" [4]. Heyting, "The Intuitionist Foundations of Mathematics" [49].

[12] Kolmogorov, "Zur Deutung der Intuitionistischen Logik" [78]. See Sundholm's paper "Constructions, Proofs, and the Meaning of the Logical Constants" [130].

[13] See Martin-Löf [90].

[14] See Kreisel "Church's Thesis: A Kind of Reducibility Axiom for Constructive Mathematics" [79]. Some of the comments below are also from Kreisel's reflections on this type of "rule" or "function" in "Which Number-theoretic Problems can be solved in Recursive Progressions on Π-paths through O" [85].

[15] Kreisel, "Foundations of Intuitionistic Logic" [80]; "Mathematical Logic" [81]. N. Goodman, [40], [41], [42], [43].

[16] Dana Scott, "Constructive Validity" [121]. Martin-Löf, "An Intuitionistic Theory of Types: Predicative Part" [91], and [92], [93]. More information on Martin-Löf's theories is presented in Beeson, *Foundations of Constructive Mathematics* [5].

Martin-Löf's work seems closest in spirit to our view, although I have not been able to make a full study of it. In [92] Martin-Löf's system uses four basic forms of *judgment* among which are the two that "S is a proposition", and "a is a proof (construction) of the proposition S". S may then be either true or false. He notes that one can read these equivalently as, respectively, "S is an intention (expectation)", and "a is a method of fulfilling (realizing) the intention (expectation) S". (Let us abbreviate this as "a:S") Then S may be either fulfillable (realizable) or not. On this reading the meaning of the logical operations can then be explained as follows: We use $\perp$ for which there is no a such that a: $\perp$. A method of fulfilling the intention

$S \wedge T$ consists of (a,b) where a:S and b:T;
S v T consists of i (a) where a:S or j(b) where b:T;
$S \rightarrow T$ consists of $(\lambda x)b(x)$ where b(a):T provided a:S;
$\forall(x)Sx$ consists of $(\lambda x)b(x)$ where b(a): Sa provided a is an individual;
$(\exists x)Sx$ consists of (a,b) where a is an individual and b:Sa;
$\neg S$ is an abbreviation of "$S \rightarrow \perp$".

Martin-Löf distinguishes canonical from non-canonical methods of fulfilling an intention and notes that this list shows methods of canonical form only. The introduction and elimination rules of the Gentzen–Prawitz intuitionistic system of natural

deduction accord nicely with the meaning of the logical constants on this interpretation.

[17] Smith's unpublished work is cited in Beeson [5], but without a bibliographical entry.

[18] See Sundholm's paper [130].

[19] ibid.

[20] A host of examples can be found in the intuitionist literature. One might see Beeson's [5] for a presentation.

CHAPTER 5

[1] Some of Kaufmann's work is now translated and published in *The Infinite in Mathematics* [74]. Gurwitsch's ideas were presented in lectures on the philosophy of mathematics at the New School for Social Research [46]. For a recent study of Husserl's texts on number see J.P. Miller, *Numbers in Presence and Absence* [95], especially Chapters 2–5. Miller comments on early and late texts on number and attempts to arrive at a unified account on the basis of these texts. The reader can find further exposition and analysis of Husserl's philosophy of arithmetic, primarily from the point of view of *PA* and other early writings, in D. Willard, *Logic and the Objectivity of Knowledge* [147], especially Chapters 2–3.

[2] *PA*, pp. 119–120. See also p. 96.

[3] See also Resnik's discussion in *Frege and the Philosophy of Mathematics* [116], pp. 43–47.

[4] *PA*, p. 116.

[5] Frege raised these and other objections in *The Foundations of Arithmetic* [34], but for specific application to Husserl's writing see Frege's "Review of Dr. E. Husserl's *Philosophy of Arithmetic*" [35].

[6] Cf. Brouwer, L.E.J., *Collected Works*, Vol. 1, [14], pp. 15, 417, 480, 509–510, 523. It seems that Brouwer made no explicit effort to avoid the charge of psychologism concerning his view of number but intuitionistic ideas concerning "idealized mathematicians" might be construed as having the same effect.

[7] Husserl had some contact with Cantor, but it is not clear what this amounted to. For Cantor's theory of number see especially "Beiträge zur Begründung der transfiniten Mengenlehre, 1", [19]. Cantor claimed that his conception of number differed from Euclid's in its replacement of the notion of "multitude" with the notion of "set". A set is not only composed of units but must also have unity. For a discussion of Cantor's theory of number see *Cantorian Set Theory and Limitation of Size*, Michael Hallett, [47], Chapter 3, especially 3.1–3.3.

[8] See *Phenomenology of Internal Time Consciousness* [67]. Section 45 is of special interest. Also, I. Miller [94].

[9] See [71], A103–A104.

[10] In [71]. A142/B182–A144/B183.

[11] See Frege's review of *Philosophie der Arithmetik* [35]. There are some other, related objections raised by Frege that I do not consider here, see pp. 320–321.

[12] In the phenomenological tradition Gurwitsch developed the view that the constitution of cardinals was founded on the constitution of ordinals in his lectures [46].

Related views about the origin of the ordinal and cardinal concepts of number have been around for a long time. See, e.g., A. Fraenkel's brief discussion in *Abstract Set Theory* [32], p. 127–128.

[13] See Husserl's *PITC* [67]. Also, I. Miller [94]. The expectation involved in protention, which concerns only immediate awareness, is part of the immediate horizon of a given act. It is to be distinguished from *acts* of expectation, of planning, etc. As we said earlier, a distinction should be drawn between the immediate horizon of an act and the global, or general horizon of an act.

[14] It is of course a theorem of elementary arithmetic that the finite ordinals just are the finite cardinals. The theorem is a *general* statement about finite ordinals and cardinals. To prove it would involve a proof by mathematical induction that two finite ordered constructions with n units are *always* ordinally similar. It seems to me that one can see intuitively that the theorem holds in specific cases involving quite small numbers. To consider the role of intuition in knowledge of the general statement we would need to show in addition that on the assumption that the statement is fulfillable for finite ordered constructions of n units it is also fulfillable in the n+1 case. Seeing that this is so involves the considerations about induction made below.

[15] It follows from our description that any fixed set of *signs*, i.e., a set of signs in which the sign to be taken as the first is fixed and so is the successor of any sign, will produce the same "natural number" if it is counted in a different order. The sequence of ordinal numbers with which it is brought into 1-to-1 correspondence will be interrupted at the same number. On sign configurations as representations of constructions see section 8 of this chapter.

[16] The exception is of course the case of very small groups of items where we can see the number of items in the group in one act, without having to run through them successively to determine their number. Since I want my remarks to generalize to larger numbers I have concentrated on the successive apprehension of items. Husserl speculated on the empirical bounds on being able to determine a number of items in one act, without having to run through them in successive acts, and evidently believed that twelve was an upper bound.

[17] It seems that Husserl had some awareness of the significance of the notion of 1-to-1 correspondence in an account of number in *PA* even though he criticized Frege's use of it in Frege's definition of number. For Husserl's criticism see *PA*. For Frege's remarks on the criticism see his review of *PA*, p. 317–318. A discussion of the dispute and of Husserl's views about the role of the notion of 1-to-1 correspondence in a phenomenological account of number can be found in J. P. Miller's book [95], pp. 58–59. I think the question of just what the difference between Husserl and Frege on this matter amounts to deserves further attention.

[18] *PA*, p. 95.

[19] *PA*, p. 102.

[20] My discussion of the nature of abstraction in this section might be compared with ideas of Nolt in his "Mathematical Intuition" [100]. I think Nolt's view of abstraction can be helpful in understanding a phenomenological view. Nolt draws some conclusions from his sketch, however, that do not seem to me to be warranted.

[21] *PA*, p. 82.

[22] Bernays "Review of Wittgenstein" [10].

[23] It should be noted, for those unfamiliar with analyses which attempt to determine

the conditions necessary for knowledge of a certain type, that there is no claim here to be describing the actual psychological stages of the mathematician at any time she happens to think about numbers. Similar misunderstandings sometimes plague those who are first reading Kant on the categories, the schematism of the understanding, synthesis, etc.

CHAPTER 6

[1] *LI*, VI, section 18. Quotes that follow are from this section.

[2] The notions of completeness and incompleteness here can be found in many places in the intuitionist literature. See, e.g., A. Troelstra's "Principles of Intuitionism", *Lecture Notes in Mathematics*, [141], where a theory of choice sequences is also described. Also, [139], [140]. Also, D. van Dalen's "Lectures on Intuitionism" [142]. The idea should also be clear from the point of view of recursion theory.

[3] E. Bishop, D. Bridges, *New Foundations of Constructive Analysis* [12], Chapter 1.

[4] See Dummett's "Wang's Paradox", [24].

[5] These comments are also meant to suggest how we might view Kant's conception of mathematical intuition. For as Kant puts it in [71] and [72], subject and predicate concepts in an analytic judgment are related solely by the logical principle of non-contradiction, while in synthetic judgments something else, some "experience", is required to connect subject and predicate concepts. In synthetic *aposteriori* judgments it is ordinary sensory experience, provided by ordinary perceptual intuition. In synthetic *apriori* judgments there must also be some actual or possible experience which connects subject and predicate concepts. This is also provided by intuition, only here it must be "pure" intuition.

In the case of either pure or empirical intuition we can view intuition as a sequence of acts in which we come to know about something. Consider Kant's example of our knowledge that $7 + 5 = 12$. The idea here is that we must undertake some acts in order to determine that $7 + 5$ is 12. We might say that the expressions "$7 + 5$" and "12" differ in sense, that they are not intentionally identical, but that in carrying out the relevant acts we come to see that $7 + 5 = 12$. That a sequence of acts is required to make the determination is even clearer when we consider large numbers, e.g., what is x where $7236 + 2729 = x$? Sensory data will not be relevant to the task of determining the number in any of these cases. The experience involved is rather simply a kind of computation. This is why Kant calls it a "pure" intuition. Only the "form" or structure of the acts involved will be relevant to the determination of number. The way in which we carry out the acts involved in making the determination will be rule-governed. The rule will fix in advance what would be given in later acts so that there is a kind of necessity here, while probabilistic elements would be involved in empirical intuition. In the case of large enough numbers, where it becomes a question of possible intuition, the importance of the *rule* comes to the fore. Kant's view would be that the rules involved here have an extra-logical content. One could not obtain them from the principles of logic.

These comments are sketchy but I believe they could be developed in some detail as an interpretation of Kant. In making these remarks I do not mean to say that Kant and Husserl were of one mind on the notion of mathematical intuition. Husserl, for

example, had a theory of intentionality, views about acts of reflection and abstraction, and views about meaning, that are not to be found in Kant. And there are ideas in Kant not to be found in Husserl.

[6] Parsons, *Mathematics in Philosophy* [107], p. 135, n. 25.

CHAPTER 7

[1] The formulation presented here is due to Charles Parsons, unpublished MS on natural numbers and finite sets [108], pp.14–15, and footnote 11, and is inspired by Zermelo's "Sur les ensembles finis et la principe de l'induction complete" [148]. We shall write $x \dotplus y$ for what is usually written $x \cup \{y\}$ and then introduce $\cup$ by primitive recursion. A well-known presentation of a theory of finite sets with individuals is Suppes, P., *Axiomatic Set Theory* [131], pp. 91–150. Suppes uses capital italic letters 'A', 'B', 'C',..., only for sets, and lower case italic letters 'x', 'y', 'z',..., which may take as their values either sets or individuals, so that the predicate "x is a set" may usually be omitted without confusion. One might also see the discussion in *Fow dations of Set Theory*, Fraenkel, Bar-Hillel, Levy, [33]. For a presentation of basic results about finite sets one might see *Basic Set Theory*, by A. Levy, [87], pp.56–63, 76–83.

[2] For example, we can define $\cup$:

$$a \cup \;= a$$
$$a \cup y = a \text{ if } \neg My$$
$$a \cup (z \dotplus y) = (a \cup z) \dotplus y.$$

We could then define, e.g., the sum set U of a set x.

[3] Discussions of the "mathematical" and "logical" conceptions of set can be found in many places in the literature. For a recent discussion which includes many of the references see Maddy, P., "Proper Classes", [89].

[4] See Bernays, P., "On Platonism in Mathematics" [11]. Basic recent sources on the iterative conception of set are Gödel's Cantor paper [39]; Wang, H. *From Mathematics to Philosophy* [145]; Schoenfield's *Mathematical Logic*, [124] Chapter 9, and [123]; "The Iterative Conception of Set" [13], by George Boolos; and Parsons' "What is the Iterative Conception of Set?" [112].

The maximal iterative conception is described by Bernays as quasi-combinatorial because it carries considerations about processes by which finite sets are constructed into the transfinite. We think, for example, of there being 2^{α} subsets of a set with α members where α is an infinite cardinal number on the analogy of there being 2^{10} subsets of a set with 10 members. Since we shall consider only finite sets it will not be necessary to take up issues about the maximal iterative conception of set or about other iterative conceptions for transfinite sets that in some way restrict the operation of going to the next stage in the process of forming sets.

One thing that could perhaps be said, however, is that concern over whether sets are predicatively or impredicatively defined is not easily set aside if we are interested in questions about intuitability. For even if we suppose that infinite processes may be actually completed, or that it is possible to synthesize infinite multitudes in a finite amount of time, there are problems about supposing that a set given by an impredica-

tive definition could be intuitable. The basic problem is this: how could a process for constructing a set be carried out if any one step in the process depends on the outcome of the process? In his comments on Russell's "vicious circle principle" in "Russell's Mathematical Logic" Gödel, for example, says that for objects that we construct there must exist a definition (i.e., a description of a construction) which does not refer to a totality to which the object defined belongs. The "construction of a thing can certainly not be based on a totality of things to which the thing to be constructed belongs". Gödel of course goes on to say that impredicative definitions are not troublesome if we view mathematics as concerned with objects that exist independently of our constructions. For then we are to think of our definitions as simply describing objects already presumed to exist, not as specifications of processes for constructing objects.

It may be that at least certain kinds of impredicative definitions are compatible with intuitability. For although the collectings performed at a certain stage must be performed on objects already given at that stage we might suppose that M can use references to future collectings (i.e., references to collectings that would be performed at later stages, hence references to sets at higher levels in the hierarchy) to individuate available entities and to collect them. Kitcher makes a suggestion like this in the context of his empirical theory of collectings in [77], pp. 145–46. However, questions about the admissability or inadmissability of impredicative definitions and of other forms of impredicativity in constructive mathematics would need to be discussed in much more detail than we can go into here.

5 *EJ*, section 61. Husserl's remark than an act of thematic apprehension is *possible at any time* may be a bit incautious.

6 This diagram is adapted from D. Willard in his *Logic and Objectivity* [147].

7 Parsons has noted this fact in his unpublished work [108], but his interpretation of what it represents is rather different from ours.

8 *EJ*, 61. It would be interesting to know whether Husserl's conception of the well-foundedness of sets influenced, was influenced by, or was independent of Zermelo's iterative conception which was first explicitly stated in his 1930 paper "Über Grenzzahlen und Mengenbereiche" [149]. It is well-known that Husserl and Zermelo had some contact in Göttingen. Some of the manuscripts on which *EJ* is based were from the years 1910–14 and other materials from which Landgrebe fashioned the drafts of *EJ* were from no later than the twenties. Landgrebe's second draft of *EJ* was completed in 1930.

9 Gödel's Cantor paper, [39], and Russell paper, [38]; Wang's [145].

10 P. Maddy, "Perception and Mathematical Intuition" [88].

11 Chihara, C., "A Gödelian Thesis Regarding Mathematical Objects: Do They Exist? And Can We Perceive Them?" [20].

12 *EJ*, sections 30–31, 62–63. Also *LI*, III.

13 In Chapter 6 of [77] Kitcher invokes the notion of "collectings", along with some other elementary operations, but with an interpretation rather different from ours. He is trying to defend a version of the view that mathematics is empirical. He also does not recognize awareness of or existence of colle*ctions*.

14 In "What is Cantor's Continuum Problem?" [39].

15 This has been emphasized by Feferman. See his "Toward Useful Type-Free Theories, I" [27], p. 80.

CHAPTER 8

[1] "Philosophical Presuppositions of Logic", in [45], pp. 350–358.
[2] *EJ*, section 64c.
[3] *FTL*, section 105. See also sections 59–60 and 106–107.
[4] On the notion of a "regulative idea" employed by Husserl the reader should compare Kant's *Critique of Pure Reason* [71], especially "The Ideal of Pure Reason", A567/B595 -- A583/B611. In these sections Kant discusses, for example, the "principle of complete determination" in connection with the "law of contradiction". In a later passage he remarks that "to know a thing completely we must know every possible predicate, and must determine it thereby, either affirmatively or negatively. The complete determination is thus a concept, which, in its totality, can never be exhibited *in concreto*. It is based upon an idea, which has its seat solely in the faculty of reason...." In saying it has its seat solely in the faculty of reason Kant, of course, means to distinguish the faculty of reason from that of understanding or sensibility. Reason, on Kant's view, always extends itself beyond understanding. Kant goes on to argue that "Although we cannot concede to (these) ideals objective reality (existence), they are not therefore to be regarded as figments of the brain; they supply reason with a standard which is indispensable to it, providing it, as they do, with a concept of that which is entirely complete in its kind, and thereby enabling it to estimate and to measure the degree and defects of the incomplete".
[5] *FTL*, sections 105.

BIBLIOGRAPHY

1. Ameriks, Karl. "Husserl's Realism", *Philosophical Review*, 86 (1977), 498–519.
2. Barwise, Jon, (ed.) *Handbook of Mathematical Logic*. Amsterdam: North Holland, 1977.
3. Becker, Oskar, *Beiträge zur phänomenologischen Begrundung der Geometrie*, 2. Auflage. Tubingen: Niemeyer, 1973.
4. Becker, Oskar, *Mathematische Existenz: Untersuchungen zur Logik und Ontologie mathematischer Phanomene*. Halle: Niemeyer, 1927.
5. Beeson, Michael. *Foundations of Constructive Mathematics*. Berlin: Springer, 1985.
6. Benacerraf, Paul. "Mathematical Truth", *Journal of Philosophy*, 70 (1973), 661–679.
7. Benacerraf, Paul. "What Numbers Could Not Be", in [8], 272–295.
8. Benacerraf, Paul, and Putnam, Hilary, (eds.) *Philosophy of Mathematics: Selected Readings*, 2nd ed. Cambridge: Cambridge University Press, 1983. Some references in the bibliography are to papers found only in the first edition of this work and will be noted as such. 1st ed., Englewood Cliffs, N.J.: Prentice-Hall, 1964.
9. Bernays, Paul. *Abhandlungen zur Philosophie der Mathematik*. Darmstadt: Wissenschaftliche Buchgesellschaft, 1976.
10. Bernays, Paul. "Comments on Ludwig Wittgenstein's *Remarks on the Foundations of Mathematics*" in the 1st edition of [8], 510–528.
11. Bernays, Paul. "On Platonism in Mathematics", in [8], 258–271. Translated by C.D. Parsons.
12. Bishop, E., and Bridges, D. *New Foundations of Constructive Analysis*. Berlin: Springer, 1985.
13. Boolos, George, "The Iterative Conception of Set", in [8], 486–503.
14. Brouwer, L.E.J. *Collected Works*, Vol. 1: *Philosophy and Intuitionistic Mathematics*, edited by Arend Heyting. Amsterdam: North-Holland, 1975.
15. Brouwer, L.E.J. "Consciousness, Philosophy, and Mathematics", in [14]. Originally published in 1948.
16. Brouwer, L.E.J. "Intuitionism and Formalism", in [14]. Originally published in 1913.
17. Brouwer, L.E.J. "On the Foundations of Mathematics", in [14]. A translation of Brouwer's thesis, 1907.
18. Brouwer, L.E.J. "Points and Spaces", in [14]. Originally published in 1953.
19. Cantor, Georg. "Beiträge zur Begründung der transfiniten Mengenlehre, I", in *Gesammelte Abhandlungen mathematischen und philosophischen Inhalts*, edited by E. Zermelo. Berlin: Springer, 1980. Originally published in 1895.
20. Chihara, Charles. "A Gödelian Thesis Regarding Mathematical Objects: Do They Exist? And Can We Perceive Them?", *Philosophical Review*, 91 (1982),

211–227.

21. Dreyfus, Hubert, ed. *Husserl, Intentionality and Cognitive Science*. Cambridge, Mass.: MIT Press, 1982.
22. Dreyfus, Hubert. *What Computers Can't Do: The Limits of Artificial Intelligence*, 2nd ed. New York: Harper and Row, 1979.
23. Dummett, Michael. *Elements of Intuitionism*. Oxford: Oxford University Press, 1977.
24. Dummett, Michael. "Wang's Paradox", in *Truth and Other Enigmas*, 248–268. Cambridge, Mass.: Harvard University Press, 1978.
25. Feferman, Solomon. "Autonomous Transfinite Progressions and the Extent of Predicative Mathematics" in *Logic, Methodology, and Philosophy of Science III*, edited by B. van Rootselaar and J.F. Staal, 121–135. Amsterdam: North-Holland, 1968.
26. Feferman, Solomon. "Some Completeness Results for Recursive Progressions of Theories (Ordinal Logics)(abstract)", *Journal of Symbolic Logic*, 24 (1959), 312.
27. Feferman, Solomon. "Toward Useful Type-Free Theories, I", *Journal of Symbolic Logic*, 49 (1984), 75–111.
28. Feferman, Solomon. "Transfinite Recursive Progressions of Axiomatic Theories", *Journal of Symbolic Logic*, 27 (1962), 259–316.
29. Føllesdal, Dagfinn. "Husserl's Notion of Noema", *Journal of Philosophy*, 66 (1969), 680–87.
30. Føllesdal, Dagfinn. "An Introduction to Phenomenology for Analytic Philosophers", in Olson, Raymond, and Paul, Anthony (eds.) *Contemporary Philosophy in Scandinavia*, 417–429. Baltimore: Johns Hopkins Press, 1972.
31. Føllesdal, Dafinn. "Phenomenology", in Carterette, E., and Friedman, M., (eds.) *Handbook of Perception I*, 377–386. San Francisco: Academic Press, 1974.
32. Fraenkel, Abraham. *Abstract Set Theory*, 4th ed., revised by A. Levy. Amsterdam: North-Holland, 1976.
33. Fraenkel, A.; Bar-Hillel, Y.; Levy, A., *Foundations of Set Theory*, 2nd revised edition. Amsterdam: North-Holland, 1973.
34. Frege, Gottlob. *The Foundations of Arithmetic: A Logico-Mathematical Inquiry into the Concept of Number*, translated by J. L. Austin, 2nd revised edition. Evanston: Northwestern University Press, 1978. This edition contains the original German text of 1884.
35. Frege, Gottlob. "Review of E. Husserl's *Philosophie der Arithmetik*", translated by E. Kluge in *Husserl: Expositions and Appraisals*, Elliston, F., and McCormick, P., (eds.), 314- 324. Notre Dame: University of Notre Dame Press, 1977. First published in *Zeitschrift fur Philosophie und Philosophische Kritik*, 103 (1894), 313–332.
36. Gandy, Robin. "Church's Thesis and Principles for Mechanisms", in Barwise, J.; Keisler, H. J.; and Kunen, K. (eds.), *Kleene Symposium*, 123–148. Amsterdam: North-Holland, 1980.
37. Gödel, Kurt. "On an Extension of Finitary Mathematics Which Has Not Yet Been Used", unpublished translation with additional notes by the author of "Über eine noch nicht benutzte Erweiterung des finiten Standpunktes",

Dialectica, 12 (1958), 280–287.
38. Gödel, Kurt. "Russell's Mathematical Logic", in [8], 447–470.
39. Gödel, Kurt. "What is Cantor's Continuum Problem?" in [8], 470–485.
40. Goodman, Nicolas. "The Arithmetic Theory of Contructions" in Mathias, A., and Rogers, H., (eds.) *Cambridge Summer School in Mathematical Logic, 1971*, Lecture Notes in Mathematics, 337, 274–298. Berlin: Springer, 1973.
41. Goodman, Nicolas. "The Faithfulness of the Interpretation of Arithmetic in the Theory of Constructions", *Journal of Symbolic Logic*, 38 (1973), 453–459.
42. Goodman, Nicolas. "A Theory of Constructions Equivalent to Arithmetic", in *Intuitionism and Proof Theory*, edited by Kino, A.; Myhill, J.; and Vesley, R.E.; 101–120. Amsterdam: North-Holland, 1970.
43. Goodman, Nicolas. "The Theory of the Gödel Functionals", *Journal of Symbolic Logic*, 41 (1976), 574–582.
44. Gottlieb, Dale. *Ontological Economy: Substitutional Quantification and Mathematics*. Oxford: Oxford University Press, 1980.
45. Gurwitsch, Aron. *Studies in Phenomenology and Psychology*. Evanston: Northwestern University Press, 1966.
46. Gurwitsch, Aron. Unpublished transcription of lectures on the philosophy of mathematics, Spring Semester, 1973, New School for Social Research. The transcription circulated among students at the New School.
47. Hallett, Michael. *Cantorian Set Theory and Limitation of Size*. Oxford: Oxford University Press, 1984.
48. Heyting, Arend. *Intuitionism, an Introduction*, 3rd ed. Amsterdam: North-Holland, 1980.
49. Heyting, Arend. "The Intuitionist Foundations of Mathematics", in [8], 52–61, translated by E. Putnam and G. Massey. Originally published in *Erkenntnis*, 2 (1931), 91–121, as part of a symposium on the foundations of mathematics.
50. Hilbert, David. "The Foundations of Mathematics", in [136], 464–479, translated by S. Bauer-Mengelberg and D. Føllesdal. Originally published in *Abhandlungen aus dem mathematischen Seminar der Hamburgischen Universität*, 6 (1928), 65–85.
51. Hilbert, David. "On the Infinite", in [136], 367–392, translated by S. Bauer-Mengelberg. Originally published in *Mathematische Annalen*, 95 (1926), 161–190.
52. Hilbert, David, and Bernays, Paul. *Grundlagen der Mathematik*, 2. Auflage. Berlin: Springer, vol. I, 1968; vol. II, 1970.
53. Husserl, Edmund. *Analysen zur Passiven Synthesis*. Husserliana, vol. 11. The Hague: Nijhoff, 1966.
54. Husserl, Edmund. *Aufsätze und Rezensionen*, Husserliana, vol. 22. The Hague: Nijhoff, 1979.
55. Husserl, Edmund. *Cartesian Meditations*, translated by Dorion Cairns. The Hague: Nijhoff, 1960. First published in 1931.
56. Husserl, Edmund. *Cartesianische Meditationen und Pariser Vortrage*. Husserliana, vol. 1. The Hague: Nijhoff, 1973.
57. Husserl, Edmund. *Ding und Raum*. Husserliana, vol. 16. The Hague: Nijhoff, 1973.
58. Husserl, Edmund. *Erfahrung und Urteil*, 4. Auflage. Hamburg: Felix Meiner,

1972.
59. Husserl, Edmund. *Experience and Judgment*, translated by J. Churchill and K. Ameriks. Evanston: Northwestern University Press, 1973. A translation of [58] which was first published posthumously in 1948.
60. Husserl, Edmund. *Formal and Transcendental Logic*, translated by Dorion Cairns. The Hague: Nijhoff, 1969. First published in 1929.
61. Husserl, Edmund. *Formale und Tranzendentale Logik.* Husserliana, vol. 17. The Hague: Nijhoff, 1974.
62. Husserl, Edmund. *Ideas: General Introduction to Pure Phenomenology*, translated by W. R. Boyce Gibson. London: Allen and Unwin, 1931. First published in 1913.
63. Husserl, Edmund. *Ideen zu einer reinen Phänomenologie und phänomenologischen Philosophie.* Erstes Buch. Husserliana, vol. 3. The Hague: Nijhoff, 1950.
64. Husserl, Edmund. *Logical Investigations*, in 2 vols., translated by J. N. Findlay. London: Routledge and Kegan Paul, 1970. Vol.I first published in 1900, vol. II in 1901.
65. Husserl, Edmund. *Logische Untersuchungen*, 5. Auflage, 3. Bde. Tübingen: Niemeyer, 1968.
66. Husserl, Edmund. *Zur Phanomenologie des inneren Zeitbewusstseins*, Husserliana, vol. 10. The Hague: Nijhoff, 1966.
67. Husserl, Edmund. *The Phenomenology of Internal Time-Consciousness*, translated by J. Churchill. Bloomington: Indiana University Press, 1973. First published in 1928.
68. Husserl, Edmund. *Philosophie der Arithmetik*, Husserliana, vol. 12. The Hague: Nijhoff, 1970.
69. Husserl, Edmund. *Studien zur Arithmetik und Geometrie*, Husserliana, vol. 21. The Hague: Nijhoff, 1983.
70. Jubien, Michael. "Ontology and Mathematical Truth", *Nous*, 11 (1977), 133–150.
71. Kant, Immanuel. *Critique of Pure Reason*, translated by Norman K. Smith. London: Macmillan, 1973.
72. Kant, Immanuel. *Prolegomena to Any Future Metaphysics*, translated by L.W. Beck. Indianapolis, Indiana: Bobbs-Merrill, 1975.
73. Kaplan, David. *Demonstratives*, photocopied, UCLA, 1977.
74. Kaufmann, F. *The Infinite in Mathematics*, translated by P. Foulkes. Dordrecht: Reidel, 1978.
75. Kitcher, Philip. "Hilbert's Epistemology", *Philosophy of Science*, 43 (1976), 99–115.
76. Kitcher, Philip. "Kant and the Foundations of Mathematics", *Philosophical Review*, 84 (1975), 23–50.
77. Kitcher, Philip. *The Nature of Mathematical Knowledge*. Oxford: Oxford University Press, 1983.
78. Kolmogorov, A. "Zur Deutung der Intuitionistischen Logik", *Mathematische Zeitschrift*, 35 (1932), 58–65.
79. Kreisel, Georg. "Church's Thesis: A Kind of Reducibility Axiom for Constructive Mathematics", in the volume cited in [83], 121–150.

80. Kreisel, Georg. "Foundations of Intuitionistic Logic", in Nagel, E.; Suppes, P.; and Tarski, A., (eds.) *Logic, Methodology and Philosophy of Science*, 198–210. Stanford, Calif.: Stanford University Press, 1962.
81. Kreisel, Georg. "Mathematical Logic", in *Lectures on Modern Mathematics III*, edited by T. L. Saaty, 95-195. New York: John Wiley and Sons, 1965.
82. Kreisel, Georg. "Ordinal Logics and the Characterization of Informal Notions of Proof", in *Proceedings of the International Congress of Mathematicians*, 289–299. Edinburgh: Cambridge University Press, 1958.
83. Kreisel, Georg. "Principles of Proof and Ordinals Implicit in Given Concepts", in *Intuitionism and Proof Theory*, edited by Kino, A.; Myhill, J.; and Vesley, R. E., 489–516. Amsterdam: North-Holland, 1970.
84. Kreisel, Georg. "A Survey of Proof Theory", *Journal of Symbolic Logic*, 33 (1968), 321–388.
85. Kreisel, Georg. "Which Number-theoretic Problems can be Solved in Recursive Progressions on Π_1-paths through 0", *Journal of Symbolic Logic*, 37 (1972), 311–334.
86. Levin, David M. *Reason and Evidence in Husserl's Phenomenology*. Evanston: Northwestern University Press, 1970.
87. Levy, Azriel. *Basic Set Theory*. Berlin: Springer, 1979.
88. Maddy, Penelope. "Perception and Mathematical Intuition", *Philosophical Review*, 89 (1980), 163–196.
89. Maddy, Penelope. "Proper Classes", *Journal of Symbolic Logic*, 48 (1983), 113–139.
90. Martin-Löf, Per. "Constructive Mathematics and Computer Programming", in Cohen, L. J.; Los, J.; Pfeiffer, H.; and Podewski, K. P. (eds.), *Logic, Methodology, and Philosphy of Science VI*, 153–179. Amsterdam: North-Holland, 1982.
91. Martin-Löf, Per. "An Intuitionistic Theory of Types: Predicative Part", in Rose, H. E., and Shepherdson, J. C. (eds.), *Logic Colloquium '73*, 73–118. Amsterdam: North-Holland, 1975.
92. Martin-Löf, Per. *Intuitionistic Type Theory*. Naples: Bibliopolis, 1984.
93. Martin-Löf, Per. *Notes on Constructive Mathematics*. Stockholm: Almqvist and Wiksell, 1970.
94. Miller, Izchak. *Husserl, Perception and Temporal Awareness*. Cambridge, Mass.: MIT Press, 1984.
95. Miller, J. P. *Numbers in Presence and Absence*. The Hague: Nijhoff, 1982.
96. Mohanty, J. N. *The Concept of Intentionality*. St. Louis: Warren H. Green, 1972.
97. Mohanty, J. N. *Husserl and Frege: A Historical and Philosophical Essay*. Bloomington: Indiana University Press, 1982.
98. Mohanty, J. N. "On Husserl's Theory of Meaning", *Southwestern Journal of Philosophy*, 5 (1974), 229–44.
99. Mohanty, J. N. *The Possibility of Transcendental Philosophy*. The Hague: Nijhoff, 1985.
100. Nolt, John. "Mathematical Intuition", *Philosophy and Phenomenological Research*, 44 (1983), 189–211.
101. Parsons, Charles. "Arithmetic and the Categories", *Topoi*, 3 (1984), 109–21.
102. Parsons, Charles. "Infinity and Kant's Conception of the 'Possibility of

Experience'", in [107], 95–109.

103. Parsons, Charles. "Intuition in Constructive Mathematics", in *Language, Mind and Logic*, edited by Jeremy Butterfield, Cambridge University Press, 1986, 211–229.
104. Parsons, Charles. "Kant's Philosophy of Arithmetic", in [107], 110–149.
105. Parsons, Charles. "Mathematical Intuition", *Proceedings of the Aristotelian Society*, 80 (1979–80), 145–168.
106. Parsons, Charles. "Mathematics, Foundations of" in *The Encyclopedia of Philosophy*, V, edited by Paul Edwards, 188–213. New York: Free Press-Macmillan, 1967.
107. Parsons, Charles. *Mathematics in Philosophy: Selected Essays*. Ithaca, N.Y.: Cornell University Press, 1983.
108. Parsons, Charles. "The Natural Numbers", unpublished manuscript of work in progress, circa 1980.
109. Parsons, Charles. "Objects and Logic", *The Monist*, 65 (1982), 491–516.
110. Parsons, Charles. "Ontology and Mathematics", in [107], 37–61.
111. Parsons, Charles. "Quine on the Philosophy of Mathematics", in [107], 176–205.
112. Parsons, Charles. "What is the Iterative Conception of Set?", in [107], 268–297.
113. Putnam, Hilary. "Mathematics Without Foundations", in *Mathematics, Matter, and Method: Philosphical Papers*, Vol. 1, 2d ed., 43–60. Cambridge: Cambridge University Press, 1979.
114. Quine, W. V. *The Roots of Reference*. La Salle, Ill.: Open Court, 1974.
115. Reid, Constance. *Hilbert*. Berlin: Springer, 1970.
116. Resnik, Michael D. *Frege and the Philosophy of Mathematics*. Ithaca, New York: Cornell University Press, 1980.
117. Resnik, Michael D. "Mathematics as a Science of Patterns: Epistemology", *Nous*, 16 (1982), 95–105.
118. Resnik, Michael D. "Mathematics as a Science of Patterns: Ontology and Reference", *Nous*, 15 (1981), 529–550.
119. Richman, F. *Constructive Mathematics*, Lecture Notes in Mathematics, 873. Berlin: Springer, 1981.
120. Schuhmann, Karl. *Husserl-Chronik*. The Hague: Nijhoff, 1977.
121. Scott, Dana. "Constructive Validity", in *Symposium on Automatic Demonstration*, Lecture Notes in Mathematics, 125, 237–275. Berlin: Springer, 1970.
122. Shapiro, Stuart. "Mathematics and Reality", *Philosophy of Science*, 50 (1983), 532–548.
123. Shoenfield, J. R. "Axioms of Set Theory" in [2], 321–344.
124. Shoenfield, J. R. *Mathematical Logic*. Reading, Mass.: Addison-Wesley, 1967.
125. Smith, David W. "Content and Context of Perception", *Synthese*, 61 (1984), 61–87.
126. Smith, David W. "Husserl on Demonstrative Sense and Reference", in [21], 193–213.
127. Smith, David W. "Indexical Sense and Reference", *Synthese*, 49 (1981), 101–127.
128. Smith, David, and McIntyre, Ronald. *Husserl and Intentionality: A Study of*

Mind, Meaning, and Language. Dordrecht: Reidel, 1982.

129. Steiner, Mark. *Mathematical Knowledge*. Ithaca, N.Y: Cornell University Press, 1975.
130. Sundholm, G. "Constructions, Proofs, and the Meaning of the Logical Constants", *Journal of Philosophical Logic*, 12 (1983), 151–172.
131. Suppes, Patrick. *Axiomatic Set Theory*. New York: Dover, 1972.
132. Tait, W. W. "Against Intuitionism: Constructive Mathematics is Part of Classical Mathematics", *Journal of Philosophical Logic*, 12 (1983), 173–195.
133. Tait, W. W. "Finitism", *Journal of Philosophy*, 78 (1981), 524–46.
134. Tieszen, Richard. "Mathematical Intuition and Husserl's Phenomenology", *Nous*, 18 (1984), 395–421.
135. Tieszen, Richard. "Phenomenology and Mathematical Knowledge", *Synthese*, 75 (1988),
136. Tragesser, Robert S. "How Mathematical Foundations All But Come About", unpublished essay, 1986.
137. Tragesser, Robert S. *Husserl and Realism in Logic and Mathematics*. Cambridge: Cambridge University Press, 1984.
138. Tragesser, Robert S. *Phenomenology and Logic*. Ithaca, N.Y.: Cornell University Press, 1977.
139. Troelstra, A. "Aspects of Constructive Mathematics", in [2], 973–1052.
140. Troelstra, A. *Metamathematical Investigation of Intuitionistic Arithmetic and Analysis*, Lecture Notes in Mathematics, 344. Berlin: Springer, 1973.
141. Troelstra, A. *Principles of Intuitionism*, Lecture Notes in Mathematics, 95. Berlin: Springer, 1969.
142. van Dalen, D. "Lectures on Intuitionism", in *Cambridge Summer School in Mathematical Logic 1971*, Lecture Notes in Mathematics, 337, 1–94. Berlin: Springer, 1973.
143. van Heijenoort, Jean, ed. *From Frege to Gödel: A Source Book in Mathematical Logic*. Cambridge, Mass.: Harvard University Press, 1967.
144. Wang, Hao. "Large Sets", in *Logic, Foundations of Mathematics, and Computability Theory*, edited by R. Butts, and J. Hintikka, 309–334. Dordrecht: Reidel, 1977.
145. Wang, Hao. *From Mathematics to Philosophy*. London: Routledge and Kegan Paul, 1974.
146. Wang, Hao. "Process and Existence in Mathematics", in *Essays on the Foundations of Mathematics*, edited by Bar-Hillel, Y.; Poznanski, E.I.J; Rabin, M.O.; and Robinson, A., 328–351. Jerusalem: Magnes Press, 1961.
147. Willard, Dallas. *Logic and the Objectivity of Knowledge*. Athens, Ohio: Ohio University Press, 1984.
148. Zermelo, Ernst. "Sur les ensembles finis et le principe de l'induction complete", *Acta Mathematica*, 32 (1909), 185–193.
149. Zermelo, Ernst. "Über Grenzzahlen und Mengenbereiche", *Fundamenta Mathematicae*, 16 1930, 29–47.

INDEX